Benjamin Coulson Robinson

Stray Thoughts on Wealth and its Sources

Benjamin Coulson Robinson

Stray Thoughts on Wealth and its Sources

ISBN/EAN: 9783337280932

Printed in Europe, USA, Canada, Australia, Japan

Cover: Foto ©berggeist007 / pixelio.de

More available books at **www.hansebooks.com**

STRAY THOUGHTS

ON

WEALTH

AND ITS SOURCES.

BY

MR. SERJEANT ROBINSON,

London:

SAMPSON LOW, MARSTON, SEARLE, & RIVINGTON,
CROWN BUILDINGS, 188, FLEET STREET.
1882.

STRAY THOUGHTS

ON

WEALTH AND ITS SOURCES.

A SKETCH.

THERE is no branch of knowledge that is less understood, less appreciated, or less cultivated than that of Political Economy. One would suppose that a science which teaches how wealth is to be acquired and increased would be a prominent one in men's thoughts and studies, since most of them are occupied with the task of augmenting their possessions and adding to their stores.

But then political economy mainly concerns itself with the mode of increasing the aggregate wealth of a nation, whilst individuals have their own particular mode of benefiting themselves, and are generally careless about the pecuniary well-being of the community, provided they themselves can manage to thrive. Individual interests are sometimes diametrically antagonistic to the general one, and men enrich themselves by methods that, if ordinarily pursued, would tend to impoverish the country. For instance, monopolies furnish means to a few of accumulating riches at the expense of the many, and every one denounces a monopolist as an invader of the rights of his fellow-men; but singularly enough when protection or its synonym, fair trade, is advocated, we find it now frequently received with acclamation by two different classes of people—those who do not understand its principle or those who hope to obtain unfair advantages to themselves by its introduction. The granting of monopolies by Charles I. as a part of his assumed prerogative, cost him his head, and however outrageous may have been such a penalty, for a mere mistake in the estimate of his personal rights, yet even in this enlightened age people cannot or will not see that what is called protection seeks to confer as grievous a monopoly as could well be devised.

Another difficulty in the way of the study of political economy is that it is supposed to involve an intimate acquaintance with statistics, and nothing can be much more repulsive than the contemplation of such a task. Then again, considering the complicated interests of mankind, the diversity of relations that subsists between the various classes of society—the theory of banking, of taxation, of stocks, shares, and exchanges—all which, it is assumed, must be thoroughly sifted and understood, it is perhaps not surprising that men's minds are not attracted to a science, in which, individually, they may deem they have no great interest, and which moreover they may think requires so much labour and devotion to accurately comprehend.

But many of these suggested difficulties are merely imaginary ones, and I venture to think that by a reduction of the matter to first principles, and by a series of plain propositions put into consecutive form, and divested of all the excrescences, by which on a first contemplation it seems to be encumbered, a better notion of the subject may be obtained than generally prevails.

I propose, therefore, to take a small community of men, locate them on an hitherto uninhabited island, and isolate them at first from all communication with the rest of the world. I, of course, furnish them at the beginning with the wherewithal to make a fair start in their new life, and then describe the course by which with proper industry and energy they may attain to considerable wealth.

I may then assume them to come into communication and competition with other communities in somewhat similar circumstances to themselves, and trace them onward to further prosperity and opulence.

But I wish it to be distinctly understood that I am not pretending to write a treatise on political economy. Neither do I arrogate to myself a claim to any fresh discovery, nor in fact to the enunciation of anything new, either in theory or argument. All I seek to do is to afford, by familiar and apt illustration of economic principles, food for thought and reflection to those who have never thought or reflected on the subject before.

Several of the axioms I lay down may be considered trite and commonplace, and scarcely worth enforcing; no doubt there are many propositions that only require to be stated to ensure universal assent; yet, taken singly and unconnected, they lead to little result. But when several kindred ones are combined, compared, and their mutual dependence upon one another made manifest, the process may be productive of very valuable conclusions. This view may be illustrated by referring to the component parts of a watch before they are put together. No one probably could look at them without being struck by the delicate minuteness and exquisite finish of the springs, the chain, the wheels, and the pinions. But it would scarcely occur to any one, without foreknowledge, that they might be so combined as to form a machine that, by ordinary care, would, with wonderful accuracy, mark the time at any hour of the day or night for twenty or thirty years.

If I should be charged with frequently repeating myself, my answer and excuse must be that whether my views are right or wrong, I am anxious at least that my meaning should be clearly understood. The same example used in different connexions may be useful in establishing many various positions.

Several of my deductions and conclusions may be entirely erroneous. These will be readily detected by those who have gone more deeply into economic science than I pretend to have done, and to their criticism and correction I freely submit myself.

But before I enter into a description of my little community, and attend it on its voyage of discovery in pursuit of prosperity and fortune, I should like to say a few words on the subject of wealth—its nature, its sources, and its different phases.. The term wealth is a very elastic, a very variable and indefinite one. The same thing may be very valuable in one condition, and utterly valueless in another. What may be of great use to one man, may be utterly useless to his neighbour. For instance, spectacles and eye-glasses, however serviceable to the many, would be of themselves valueless to the blind. One who could not read might just as well be without books as possess them. A merchant cast with a quantity of costly merchandise on a desert shore would be none the richer for his surroundings : he might perish for lack of nourishment in the midst of what, under ordinary circumstances, would be regarded as enormous wealth.

Again, one who had a much larger stock of provisions on his hands than he could possibly consume would, if he had no clothing, die in the midst of a rigorous winter from cold. Another who had a vast superfluity of clothing would, if he were without food, starve from hunger.

Now this evidently introduces into the definition of wealth, the attribute of exchangeability, or the power of parting with what we do not want for that which we may require, or wish to possess. For example, in the instances I have just put—if each man could have parted with his surplus to the other, both might have lived and prospered.

The sources of enjoyment are manifold, and they operate differently on different individuals; and, moreover, there is this tendency in most of us, that what we have once vehemently desired, soon palls upon us when obtained, and we sigh for some new and untried means of gratification.

The consequence of this is, that among all mankind, the peer and the peasant, the rich and the poor, there is a constant, never-ending traffic carried on—a continuous interchange of commodities taking place, according to the requirements or the necessities of the various individuals. This exchanging is effected by means of money, though the articles purchased by it are the main objects of desire.

Wealth then may be safely defined to be that which gives its possessor a command over the *necessaries*, the *conveniences*, and the *luxuries* of life; but unless he had the power and opportunity of parting with what he does not require, for that which he may wish for, a large portion of his possessions become valueless to him. We see then that the principle of exchangeability is clearly included in the term Wealth.

By *necessaries* I mean what is absolutely essential to man's continued existence, and they comprise in their most simple form, food, clothing, and shelter.

Conveniences include what, though not actually necessary to us, yet still contribute materially to our comfort, or to the means of adding to our possessions; such as, for instance, chairs and tables—tools which save us much manual labour, carts and carriages which save both labour and time.

We come then to *luxuries*, which may be taken to be those things which tend merely to the gratification of the senses, the appetite, or the taste, without the slightest other real utility to the possessor or the world, except that the desire to possess them may stimulate people to work and labour, and so contribute to the substantial wealth of the community. I use the word substantial in a modified sense, because although works of art, for instance, may benefit a nation by civilizing and refining it—however they may enrich the individual who executes them—they do not in the slightest degree add anything to the productive capabilities of the general society. They have an exchangeable value, if a market can be found for them; but however many hands they may pass through, neither they nor the thing which is exchanged for them involve any increase in their respective worth.

Among luxuries (merely to give examples) may

be classed such things as precious stones, sculptures, paintings—as well as expensive wines and other objects which merely please the palate, as distinguished from those which contribute to the sustenance of the individual. It may be mentioned, however, here that many of the things usually considered as portions of a man's wealth may partake of the character of both conveniences and luxuries. For instance, a carved ebony table would not be really more useful than a common deal one. A richly chased goblet would not be more useful to drink out of than an ordinary delft one. As far as mere utility went they would be upon an equality; yet there might be an enormous difference in their respective money values, where there was an open market for their disposal. But the article and the money paid for it would merely change hands, and there would be no more real value in the community than before. To possess the superior piece of workmanship might please a sentiment or a taste; but in the possession of one, or of a succession of owners, it would do no more. Assuming it to be an heir-loom, which could not be parted with, it would only remain to be looked at, and hundreds of people might have the same pleasure in that respect as the proprietor himself.

In the same way it may be said that our open

parks and museums afford as much amusement and pleasure to the public as to those who own them. They could scarcely be said to be the source of wealth to those whose actual property they might be. But call the right or title what we will, this consideration illustrates the difference between marketable and unmarketable value.

A wealthy man then means one who has more or less extensive property, a large portion of which he does not require for his own use, but which he has the means of exchanging for other property, or means of enjoyment, that he has a necessity or a fancy for.

It is said that none of us want to part with our money, but yet we are hourly and daily spending it with the view of obtaining more gratification than we should get by its retention in our purses.

If we were to keep it perpetually locked up in a chest—for all available service it would be to us—it might just as well be at the bottom of the sea.

But the fact must never be lost sight of that this bartering and exchanging, this buying and selling, add nothing, in themselves, to the wealth of a community. They stimulate production by inducing some to produce what others may require. They constitute the machinery by which wealth may be distributed and indirectly augmented; but, as I have said, they add nothing to the value of the articles exchanged, since what one man gains, the other loses.

With regard to the relation of money to wealth, I shall dwell on this subsequently; but I may say here that all buying and selling is mere bartering one thing for another. If you buy corn with money, you buy money with the corn; and the one is just as much a commodity as the other. However obvious this truth may be when stated, it is a much more important proposition than may at first appear, as I hope may be seen hereafter.

It is with nations precisely as it is with individuals: trade and commerce, *per se*, have no effect in increasing the value of the goods exchanged. Still, one country may have peculiar facilities for yielding or forming one species of things, while another may be singularly prolific in developing a different kind of produce which the first may require. The result, where there is unrestricted intercourse, would naturally be that each would produce more than it required for its own consumption, of what it could so advantageously create, in order that it might obtain from the other that of which that other had a profusion, and though there would be no increase in either of the commodities exchanged, the transactions might very much tend to furnish a means of increase, as we shall see by and by.

But in any way it is easy to perceive that the two countries would be largely benefited by each obtaining what it might otherwise be in need of, either in the shape of necessaries, conveniences, or luxuries.

What, then, is the real origin of the world's wealth, viewing it with reference to the progressive increase of property and possessions from year to year. It is palpable that there is an enormous consumption and waste of valuable produce going on continuously. Yet, notwithstanding this destruction, we possess infinitely more riches than existed in the earlier stages of mankind.

Since the creation it might perhaps startle us to find that successive generations of men and animals have consumed and destroyed as much in bulk as would equal the substance of the earth itself. Of course, when I use the word destroyed, I do not mean it as an equivalent to "annihilated," because we know that no particle of matter, however it may be changed in form, is really lost to us. I only mean that what was once wealth, and had a specific value, is changed into something valueless and useless, commercially or otherwise, although by the combined operations of nature and the energies of man the most worthless and noxious things may form the elements of future utility and wealth, and so contribute to the necessities, the comforts, and even become the source of luxuries to mankind.

Now if this vast increase in opulence may be truly asserted, how has it been brought about? Evidently by that part of created things that possesses vitality, and is thus capable of reproduction. Rocks, stones, the solid earth will not increase and grow. Neither a pebble nor a diamond can multiply itself; they will remain the same for all time, except by mechanical or chemical action in some way exercised towards them. Their form may be changed, but they cannot reproduce themselves. But all things that have life—men and other animals, plants, trees, vegetables—have the faculty of propagating themselves, and are capable of reproducing a large number of their own species, and it is worthy of remark, as evidence of providential design, that the lowest classes in the scale of vitality—those which are subservient to the wants and necessities of the higher ones—have in general by far the greater power of reproducing their kind, so that there may be less chance of failure in that which may be essential to the existence of the superior grades. For instance, there is a limit, though an uncertain and undefined one, to the procreation of children, both as to number and time. That limit is far less applicable in the case of most domestic animals, while the seeds of plants and fruit will reproduce at the rate of a hundredfold.

We may then safely assume that the condition of vitality is the primary source of that which produces the increased wealth of mankind. There can be no other principle or example of increase—of the recuperation of that which has been destroyed. It is pretty clear, too, that men were intended from the beginning to work and toil for their subsistence. Omnipotence might, if it had so pleased, have caused the earth spontaneously to bring forth all the necessaries and luxuries that human beings might require for their existence or enjoyment, and in such abundance that they might have them for the mere seeking: but then there would be no scope for the employment of those faculties with which they are endowed, and which must have been implanted in them for some special purpose.

Men differ essentially from the lower animals in this respect, that whatever is necessary for the continued well-being of the latter, and for the fulfilment of their destiny, was provided for them with a lavish hand in the spontaneous and luxuriant vegetation with which the particular region in which they were placed abounded. They were not gifted in general, as superior intellects are, with foresight to perceive that their present store of food might be exhausted, and that it might be necessary to provide in time for its renewal. What they required was plentifully supplied to them by the teeming woods and plains, and by the copious streams that rushed down from the hills. There was no need of cultivation, nor for any labour, to ensure a supply of what their appetites might demand.

Moreover, they were given natural instincts which sufficiently instructed them in choosing that species of food which was best adapted to their peculiar physical conformation, with reference to health and longevity.

Nature, too, without any exertion of their own, gave them, according to circumstances, warmer clothing in the shape of wool and fur in winter, while the foliage of spreading trees afforded them shelter from the piercing rays of the sun in summer.

There is again this striking peculiarity about them, that the food which they eat and drink, their means of enjoyment, their sources of pleasure and pain—in fact, their culture and civilization, if I may use the expression—were in general precisely the same probably at the creation as they are now.

This may be fairly accounted for by the fact that they have no language, and this circumstance would deprive them of all means of organization for the improvement of their condition, even if it were otherwise susceptible of advancement. No doubt they have limited means of communication with one another—of the nature of which we know nothing—but it merely resolves itself into one of those instincts which were bestowed upon them for the purpose of satisfying their immediate needs. Everything tends to show that the lower animals were intended to be subservient to the wants and requirements of man.

Probably in the very early period of the world men too were in much the same condition as the brutes in regard to their means of subsistence; they lived at first on berries and the wild animals they killed, and whose skins supplied them with clothing, but they were imbued with higher aspirations, and were endowed with the desires and the means of augmenting the enjoyments and comforts of life, and they would thus make speedy progress towards comparative civilization.

Now assuming that it was part of the original

design that man should toil and labour, it is difficult to imagine a more effectual mode of carrying this out than by rendering those things that were most necessary to his actual existence speedily perishable. Meat, bread, fruit, vegetables, very rapidly decay, and perpetual labour and exertion are required to reproduce them. We cannot eat corn nor feed upon cattle as such, though we do partially live on the product which each furnishes. It is true, corn can be stored, fruit may be preserved, and cattle may live for a time; but even then they require labour to prepare, keep, and properly tend them.

It is, however, enough for my purpose to point out that the most pressing of our necessities, namely, food in general, decays more quickly than other things which, though essential to our well-being, are not so indispensable as the former. It would be obviously useless therefore to produce at any one time more perishable food than could be consumed within a given period, and thus arises the necessity for never-ceasing labour, in order to secure a continuous supply. It may be urged that only a few people, comparatively, are employed in producing the actual necessaries of life. But it must be remembered that the rest must resort to toil and exertion in order to produce other things which may tempt the cultivators of the soil to part with a portion of their food in exchange for them.

Now it may fairly be asked—if what we call wealth depends on the continuous reproduction of what in itself is perishable, and therefore cannot be long hoarded up, by what process do we go on constantly increasing it? The answer is simple. All realized property is in truth mere stored up labour, or, in other words, it represents the food and other necessaries that the labourer has consumed in producing or manufacturing it. The food has been used up, but it survives in the thews and sinews of the worker; his labour is exhausted, but it survives in the article, whatever it may be, upon which that labour has been expended.

The result of the operation then is this, the labourer is in the same, probably in a better, condition than when he began, and he is ready to begin again. He has been nourished and supported during the time, by the necessaries (or what is the same thing, money to purchase them) with which his employer has supplied him; while the employer has exchanged those necessaries which were perishable, and were useless to him personally, for a thing which would be more or less durable and permanent, and which must of course tend to his gratification, or he would not have entered into the transaction. And this in its simplest form may serve as an exemplification of the principle on which wealth is founded.

If mankind could exist without food and other necessaries, and men still retained their energy and habits of industry, they could only employ their time in forming articles of convenience and luxury, each using his skill in producing what might be enjoyable by others, in order that he might exchange it for those products of theirs that might afford pleasure to *him*, and so there would be an ever-increasing accumulation of property as time went on.

On the contrary, if, constituted as we are—compelled to toil for our subsistence, the soil were so churlish as, with all our exertions, to produce no more than was sufficient to furnish us with the actual necessaries of life, upon which employment every one must be engaged, there would be neither time nor opportunity for the creation of any adjuncts to our comfort or our luxury, and there could be no accumulation of what is called property.

But we are subject to neither of these conditions. The earth is so teeming and prolific that if the need existed and other occupations were relinquished, we might raise infinitely more of what constitutes the mere means of living than we do now.

Besides, what one country cannot produce easily in the shape of aliment, another can, and this suggests the immense advantage of a free intercourse among the various nations of the globe. Cattle are slaughtered in Australia and in South America for the sake of their hides, or fat to be used as tallow, and vast quantities of meat are wasted, which might otherwise serve as sustenance for starving populations in other parts of the world. Corn grows in Western America in wonderful abundance, almost spontaneously, or at least with the exercise of very little manual labour, and the ingenuity of man is taxed to find means of profitably transporting it to distant regions which require it—such as ours, where the available soil has become less fertile, and is in a measure exhausted.

At all events the fecundity of the earth generally is so great that a comparatively small section of mankind may suffice to furnish actual necessaries for all the rest. These latter are therefore set free to shape and form other products of fixed and otherwise non-generative materials into what may be useful or material to mankind. Clay may be turned into bricks, bricks into houses; wood may be shaped into furniture; stone into monuments and temples. Iron may be made into tools and implements of all kinds. All these would be more or less durable, and while in that condition they would in fact constitute the wealth of each succeeding generation. To sum up the argument, Nature supplies us almost gratuitously, as far as the exertions of all mankind are concerned, with the necessaries of life. These cannot be stored in specie, because they are, in general, perishable; but they can be stored in the result by the products of the labour which this boon of nature supports and nourishes during the period of their formation.

The money value of a thing roughly stated must ordinarily be regulated by the quantity of labour, or, in other words, the quantity of food and other necessaries consumed by the labourer while producing it; but it is not affected, as is often erroneously supposed, by any temporary variation in the price of that labour. Wages depend in general on a totally different principle, namely, on the amount of capital in proportion to the quantity of labour in the market, as will be seen further on. Whether more or less is paid for that labour does not in the slightest degree affect the money price of goods.

But of course there may be fancy prices, which have nothing to do with the mere cost of production, but depend on the rarity of the particular article, or the difficulty and perhaps impossibility by any amount of toil of multiplying it. Precious stones, old coins, ancient statuary, valuable by reason of their scarcity or antiquity, may be mentioned as examples of the class.

Another matter in connexion with this subject it may be important to refer to.

It by no means follows that those things which are most useful, and even essential, are most valuable in a monetary point of view. Otherwise air and water would be the most costly of all products, for we could only live for a few minutes without the one, and but for a short time without the other, and this again illustrates the difference between interchangeable or commercial value, and one which, though supremely important in our vital economy, cannot in general be estimated in figures or in coin.

To render a resort to any such standard available, we need a commodity of necessity or utility, of which some people have a superabundance and others a deficiency, and this can only be said of things of which, taking the whole community together, there is a limited supply.

Now air and water are among the exceptions to the rule I have before referred to, that those things which are most essential to the continuance of our being we must needs toil and labour to obtain.

But these two requisites, among others, are given to us in such profusion, under ordinary circumstances, that no one can require to prepare them for himself, or to buy them from any one else. It is true that air may become vitiated, or water may be dangerously scarce in particular localities. Purer air or a supply of water may then command a price, but it would mainly consist of the labour spent in remedying these evils. The sand of the sea-shore we look upon as almost valueless where it is originally cast, but if it is required in the midland counties it may have there a considerable money value, and that would mean the cost of the labour of collecting and conveying it across the country.

But I am only dealing with general principles, and unless I desire to illustrate a particular position I trouble myself little about exceptions, although there are few general principles without them.

How is it then that we get air, for instance, in such abundance without any exertion of our own.

In the first place, had we to provide air for ourselves, the slightest delay in the operation would be fatal.

We may live for a long time without food, and during that period we are warned to seek it. In the other case, a few seconds might constitute the difference between life and death.

Again, air and water are as necessary to the brute creation as they are to mankind, and as I have endeavoured to show, Providence has bestowed actual necessaries much more lavishly and spontaneously upon them than upon man.

Moreover, air and water exercise vast influence over most things, both animate and inanimate in the creation. We take them as special gifts bestowed equally on rich and poor, and which we have no need to waste our energies in procuring for ourselves.

I have already used the word capital, and although it will be more fully referred to and explained further on, it is right that I should expend a few words upon it here.

However exuberant and fertile a soil may be, Nature requires time for her operations to be carried out.

If we could get corn the day after we had sown the seed, or if a sapling could give fruit the day after it was planted, it is clear that the law which necessitates a resort to labour would be partially annulled.

The seed may reproduce itself manifold, but before it comes to maturity much labour must be expended in its cultivation, and much time must be consumed before it is developed into food. During this time those who cultivate it must be nourished and supported, and this can only be from that which was previously realized, saved and stored in the past, in the shape of food or of other property in exchange for which food, &c., may be obtained.

Then there is the mill, the plant, the machinery, the tools. All these must have been the result of thrift and the storing up of what remained unconsumed of past production.

Now that which goes to support and maintain the labourer or workman while he is bringing his work to perfection is called capital; and the more of it that exists in any particular country, the richer of course in general will the community be, and the better off will be the working classes who are fed and supported by it. Nothing can be more mendacious than the assertion that there is a natural antagonism between capital and labour. There are always bad advisers in the world, who find their account in flattering the worst passions of the multitude, and in leading them astray from the path of contented industry.

It may be that capitalists as well as workmen may sometimes seek to obtain a larger share of their joint gain than is quite consistent with equity; but that unfair attempt would soon correct itself. The relative proportion between the profits of the capitalist and the wages of the labourer is regulated by a law far beyond the control of either master or man. I have before said that the money value of the article manufactured is not in geneal affected by either, but that price is fixed according to the quantum of labour bestowed upon it, and not upon the sum that is paid for it. Price is ruled by a law of its own; profits and wages have one between them, and it decrees that each shall depend upon the relation of the amount of capital to the amount of labour in the market. Competition would soon repress any transgression of nature's ordinances in this respect.

There is no better established proposition in political economy than this, that the wages of labour and the profits of capital come out of one single fund which consists of the surplus of sale value left, after all *other* expense and outlay of production have been allowed for. The price of the manufactured goods depends mainly upon that expense and outlay. The surplus—a totally different thing—is to be appropriated exclusively to the profits of capital and to wages. The question is how is this to be divided between them. The solution of this problem entirely depends, as the money value of most other things depends, upon demand and supply. It is clear that when there is a larger quantity of goods in the market than the public exigencies require, the price of those goods will fall. If, on the contrary, there is less than can accommodate the normal wants of individuals, their market value will rise. Now capital and labour are in this respect subject to precisely the same conditions. If the labour market is suddenly overstocked in relation to the capital that is to employ it, there are more labourers contending for the same aggregate amount of that capital in the shape of wages than before, and they will compete with one another by reducing their demands for remuneration. Assume, on the other hand, that capital has largely increased, with fewer labourers to bid for it, there must necessarily be a larger competition amongst the capitalists for labour, wages will inevitably rise, and the profits of the capitalists will as surely be reduced. If a certain fixed sum is to be divided between two parties, where one of them gets more, the other must obviously get less. The proof that an increase of wages must come out of profits, and reduce them, may be a little diversified in this way. I assume that where there is a glut of anything in the market, that is, where there are more sellers than buyers, the money price of that thing must fall; it becomes of less value to its particular owner, that is to say, his profit on the sale is lessened. Just in the same way, when there is a large redundance of capital, it becomes of less value, in the sense that it is less productive to its possessors. Each of them has a greater amount of competition to contend against than when capital was scarcer, and he must needs embark his funds in less profitable investments than he did before. Just as in a crowded room every one must content himself with a more limited space than when it was only moderately full.

If then it be conceded that profits fall as wages rise, and *vice versâ*, and that any variation must result from a change in the ratio that the amount of capital bears to the supply of labour, it follows that profits and wages will continue much the same as long as that ratio continues, and that no individual of either class could have any power to alter it.

If an employer sought arbitrarily to reduce wages, that he might augment his profits, and was really successful in his attempt for a time, other floating capital, of which there is always a vast amount ready to be invested in any speculation that is likely to turn out extra remunerative, would at once rush into the particular trade, and by competition speedily reduce his profits to their normal level. He would soon discover how futile had been the attempt, and his regret at his folly would not be mitigated by finding that he had been the means of introducing into that trade a large number of competitors who would manage to divide with him in future the legitimate gains that were once all his own.

A kindred fallacy pervades the minds of many who only regard these matters superficially. It is often assumed that if manufacturers were to pay higher wages to their workmen, they might easily recoup themselves for the difference, by adding it on to the price of their goods; but I trust I have already shown that the sellers of this kind of property have substantially no more power of deciding what shall be their selling price than their customers have.

There is always—at any one period of time—a

fixed and recognized rate of the profits of capital, and an equally settled rate of the wages of workmen, varying, of course, with the amount of skill, experience, or intelligence required in any particular trade or occupation; and any attempt to interfere with these would, as I have said before, be simply to invite a ruinous competition. Employers will always take care not to give a higher rate of wages than the current one, or it would interfere with the sale of their goods if they were bent on reaping the current profit.

If they could put an increased price on goods to compensate for increased wages, they might as well add a further sum for the purpose of augmenting their own gains. The inevitable result would be that they would be undersold by their rivals, and their stock would remain upon their hands until they had arrived at a sounder appreciation of their own interests.

What would be the result of all the grocers in a particular locality combining and agreeing to put a higher price upon their wares than the ordinary one. Why, that enterprising grocers from distant neighbourhoods would invade their domain, and by underselling them, speedily take away their custom.

If then a few manufacturers or producers were to put an exorbitant price upon their productions, we see how speedily competition would defeat them. If it were possible to conceive such an unanimity in unreason that all should combine to sell their goods at an unnecessarily dear rate, the price of all merchandise would of course be increased; the result of this would be that the incomes of individuals would not bear the strain, a less quantity of goods would be purchased, and the manufacturer would be no better off than before. If he gained on each individual transaction he would lose by the number of transactions being reduced.

Take the case of the omnibus and steam-boat companies. At the fares they charge now they make an infinitely larger amount of profit than they made when the charge was much higher. Multitudes of people use this mode of transit now who never thought of using it before. Were the directors again to increase the fares to a threefold sum they would probably reduce the number of passengers to one tenth of what it is now.

I hope then I have made it manifest that the outcry on the part of working men against capital in general is about as suicidal a one as they could adopt. Even if they were the victims, temporarily, of some injustice, the worst possible mode of resisting it would be to resort to those violent measures which are often recommended and even forced upon them by their factious, interested advisers. Strikes and turn-outs are infinitely more disastrous in their results to the workmen than they are to their employers. The latter suffer, no doubt, by the loss of the return which their invested funds ought to afford them. Their capital becomes in the end lessened; they must live upon it. Consumption is still going on, and there is no reproduction of that which is consumed; and in this way the labourer also suffers indirectly, for the funds which were to supply him with work are so far reduced.

But he of course injures himself still more by being compelled to consume his savings in supporting himself during the time he remains idle, or he sacrifices his independence by living on the miserable pittance doled out to him by the charitable contributions of others who are more prosperous, though equally misguided with himself. It may be necessary sometimes to withstand the sordid attempts of employers to reduce wages below their proper level, but strikes are much more frequently entered upon at the dictation of one of the working classes themselves, who lives at ease upon their alleged grievances, and profits by the misery he is continually lamenting.

I have said that the mission of man in this world is to toil and labour for his subsistence. But it may be fairly urged that if this is so, the behests of Providence have not been carried out in the case of people with large fortunes, who are thus supposed to rebel against Divine ordinances. It is this sort of feeling that begets in the toiler an idea that some people are favoured by Providence in the distribution of the good things of this life, and that wealth is often bestowed where it is least deserved —that, in fact, the poor are compelled to labour to support the rich in the indulgence of their pleasures and their fancies; and this further absurd deduction is often made by those who know no better, though frequently encouraged to make it by those who do—that what they conceive to be the charms and delights of life are entirely created by their labour, and that they ought to be considered the sole and actual producers of that which confers upon the rich their various means of enjoyment. Just as well might the masons who sawed and placed the blocks of stone upon one another at the building of St. Paul's, claim the credit of having erected that magnificent pile; or a working tailor to whom cloth was delivered to be turned into a suit of clothes, claim the suit when completed, on the ground that it was his handiwork. It is true that the mason and the tailor were necessary to produce the respective results, but where would they have been without the capitalist who supplied the materials and their subsistence during the operation. At the risk of repetition, I must remind the reader that the capitalist who employs a workman to exercise his calling and his skill for wages in the manufacture of any articles, loses the value of the necessaries with which he has supplied the artificer while engaged on his work; *that* property or value is utterly lost to him, but he gets for it in exchange what he of course considers more than an equivalent; the workman gets his share of the advantage of the transaction in the wages that are paid to him; and he deems himself benefited by the bargain, or he would not have entered into it. True, the owner gets the article when made, but then he had the value of it at the commencement of the contract, and must necessarily be entitled to retain it at the end. The workman owned nothing except his prospective labour; why should he be better off, beyond the wages that paid him for that labour? It is he that has eaten up and consumed the amount of capital with which the employer has furnished him. Surely it is but just and reasonable that he should replace it. Had it not been for this little store which the employer possessed, he might have been without work and without food.

Conceive a state of things where no capital of any kind existed; where nothing had been saved or stored by any individual from the bounteous products, or the results of them, that Providence had lavished on mankind. There could be no employment of one man by another, for there would be nothing wherewith to remunerate the *employé*. It is true that one species of labour might be contracted for in exchange for another; but this could be but an uncertain promise, to be fulfilled in future, and might be broken in the end. Capital, or the funds saved by former industry and frugality, is the very foundation of the comfort and prosperity of the labouring man who begins the world with nothing. He ought to look upon it as his greatest blessing, instead of regarding it with suspicion and distrust, and seeking to diminish it with a sort of half-concealed, foolish instinct of wishing to reduce its possessors to a level with himself. In envying what he assumes to be the happiness and prosperity of the wealthy, he loses sight of the fact that the more riches men have accumulated, the larger is the share of it that he secures to himself. It must never be forgotten that large possessions which enable a man to remain idle, or at least, assure to him an immunity from manual toil, are after all mere hoarded labour, which really means the hoarded food and other necessaries that have supported the worker during the time that he was completing the subject-matter of this wealth. Its possessor or those from whom he legitimately derived it must by their heads or their hands, by their industry and their thrift have stored up for future use, more of the aggregate products of the soil or that which would purchase them, than others of their fellows had done. They might, in so doing, have acted selfishly, with a view of exempting themselves at a future period from the necessity of working, or perhaps with the natural desire of placing their progeny in the same position; but whatever the motive might be, there can be no doubt that it would tend materially to the advantage of the whole community. They might have been content with satisfying their own needs or their pleasures for the moment. They might have yielded themselves up to a life of ease and luxury, and have been satisfied to die, leaving no property behind them. But they chose to work on, and by continued perseverance and industry, to add yearly to their stores a large amount of wealth which would serve not only to benefit their descendants, but would in ordinary course almost necessarily be employed in the support of future labour.

Those who view these things without bestowing much thought beyond the surface, are wont to assume that a millionaire is the sole enjoyer of the property he possesses; and that by the expenditure of his income on his own pleasures, so much is wasted of the aggregate wealth of the country. There cannot be a more egregious fallacy. However wealthy a man may be, he cannot consume personally a larger amount of actual necessaries than one of his own dependants; and remember that what he does consume is the only actual loss the community sustains through him. The rest of his income is consumed or eaten up by those whom he employs in various capacities. It is true that what he eats and drinks and wears may be richer and more expensive in quality; but the excess would be very trifling amid the general expenditure of capital or income; and even here some people must be employed, nourished, and supported by his funds, in providing this very excess.

Let me assume that a man has a spending power

equal to 100,000*l.* a year, and then trace how, according to probabilities, he would dispose of it. The chances are that it would all be paid away in the course of a year to agents, tradesmen, servants, and workmen of all kinds; and each of these would, in his turn, pay away his earnings to others, whose goods or services he requires; these again would do the same thing, and so on. Every one would gain something to himself out of that large fortune, and thus a multitude of persons would be supplied with employment (which means nourishment and comforts) by the accumulated wealth of one individual.

If he were a prudent man, he would so utilize his property as to secure the same income for every succeeding year, and this could only be brought about by the process of reproduction, the result of the labour of those whom he employed. But if he were recouped for all he had laid out or paid away, the same might be said of all the numerous people through whose hands the money had filtered; they too had been maintained probably in comfort during the year, and would find themselves at the end of it at least as well off as they were at the beginning.

But let us suppose that this wealthy capitalist was not a prudent man, and we shall find that there would be very little difference in the result. We have heard of late years of persons with a rent-roll or an income far exceeding the figure I have above mentioned, who, by their recklessness, extravagance, and luxury have frittered away their fortunes, until they have become bankrupts and beggars, and a foolish outcry has been raised against them that they have injured the public by wasting and annihilating funds which ought to have gone to the support of honest industry. They may be justly blamed and denounced for their folly, but they are probably the only sufferers, and it may be that the substance of their former wealth is as existent and as great as ever; the only difference is that it has got into other hands, and possibly into such as will make a much better use of it; but of this, the public, at all events, have no reason to complain.

Of course there is a great difference between what is called productive and unproductive labour. Digging a deep hole in the ground, with no other object than that of filling it up again, is an instance of the latter. It is simply a waste of time and of food, for the labourer must eat and drink while he does it; but the toil of sinking a well in a parched soil is productive labour, for it survives in the means it affords of irrigating the earth, and rendering it perhaps much more fertile in the future. All these principles will, I hope, be further illustrated when I come to speak of my little colony, and start it into life and action. This I now propose to enter upon.

But in doing so I must be allowed to make many assumptions, and to lay down many conditions, before I send it forth to try its fortune in a hitherto unknown land, and I want its precise position to be clearly understood from the beginning, so that the change from what it then was, to that to which it might eventually be brought, may be more easily traced and followed out. It must be stated that the numbers and figures, whether of men or things with reference to my general illustrations, are quite arbitrary, and have no regard to the due proportion between them and other matters with which they may be connected. I wish to dwell on natural laws and principles—quite regardless of facts—still less do I predict inevitable results. I am only desirous of reducing the subject to its simplest elements, and to rid it of all extraneous topics which might otherwise hamper and complicate its consideration.

I will suppose, then, twenty men of average intelligence, but with varied capacities, located upon an island, and far removed from any communication with the outer world. The new territory shall be of extent sufficient, when fairly apportioned amongst them, to give ample scope for the exercise of their energies and faculties, in their endeavours to improve their condition. It shall be sufficiently fertile to produce by cultivation and industry not only whatever may be necessary for the subsistence and well-being of the new-comers, but also what may contribute to the conveniences, comforts, and pleasures of life.

I mean, therefore, at the outset, to give them wherewithal to supply not only their immediate necessities, but also what would be requisite for their future subsistence during the time that their efforts for providing for themselves were maturing. I have already partially referred to that stage of the world's career when men were in the condition of aborigines and had to seek nourishment from the spontaneous productions of the soil. I wish it to be understood that my small society has had all the advantages of civilization, and that they go forth as emigrants constantly do, to tempt fortune in a new and untried land. And I may add that I leave out at present all reference to an increase of population, because family considerations would interfere with the simplicity of a mere abstract discussion as to how wealth may be promoted and maintained.

I propose, therefore, endowing each of the emigrants with property which would cost in the country they had left the sum of 100*l*., so that they all start on equally fair conditions. Their future prosperity would much depend upon how each 100*l*. had been laid out. I need hardly say it would be absurd to take any large portion of it in money. There would be no one to buy of, except their companions, and this would be a mere exchange, adding nothing to the joint stock of the company.

Costly furniture, expensive household utensils, jewellery, or other mere ornaments would of course be of no avail, if forming any portion of the 100*l*. outlay; they would be absorbing a portion of the money that might be much better employed. Their stock should mainly consist of that which, though more or less consumable, might with care and industry be reproduced.

Food would of course be the first consideration. Seeds, plants, a few cattle, sheep and poultry, two or three horses and mares; clothes to cover them; materials for building huts to shelter them, tools, implements, &c., all these would be essential, for though these latter are not reproductive of themselves, they would materially aid in reproduction. But everything they invested money in should be the plainest and cheapest of its kind, consistently with use and service for the future.

A portion of their capital (for this would be its proper designation) must necessarily consist of money, by which I mean coin. I need scarcely suggest that a bank-note or bill, however useful elsewhere, would, if it were taken as a part of the 100*l.*, be of no possible value here. It would be mere waste paper to its possessor, unless the bill were drawn on one of his comrades, and if he paid it, his 100*l.* worth of property would be lessened by precisely the amount of the bill.

Notes and bills are serviceable by providing an easy mode of transferring debts. They are the mere shadowy representatives of a certain amount of coin in other people's pockets, but the value of that coin is assuredly not doubled by writing or printing anything on a worthless piece of paper.

In general, a system of paper currency is regarded as an indication of poverty on the part of the nation that resorts to it extensively. It suggests that that nation cannot afford to purchase the metal that would be requisite to turn into coin, and is obliged to adopt a cheaper mode of providing for the circulation of wealth.

And at all events in a highly-taxed country it

might not be expedient or safe to impose further imposts upon the population. Paper is cheaper than gold and silver, and where a paper currency is general, it is made compulsory on the inhabitants to receive it in payment of their debts. It is, however, for the most part at a discount, that is to say, its interchangeable value is less than its nominal one, and the reason is obvious; people prefer to have cash in hand to trusting to the indebtedness of others, however respectable or responsible those others may be, and although the most unstable governments usually guarantee the ultimate payment of their notes, they almost invariably omit to say when the payment shall be made. But government guarantees are not always to be relied on. Foreigners cannot, at all events, be expected to put full confidence in them. It is true, with us in England, bank-notes are always as valuable as coin, but then their issue is subject to great restrictions, and an amount of substantial bullion, largely proportioned to that issue, is always kept ready at hand to discharge the debt when payment is demanded, so that ample faith in the contract is established. It may be said that, such is the trust reposed in those who issue Bank of England notes, that even throughout most of the states of Europe they generally realize the same amount as the current coin of the realm. But it is enough for my purpose to say that whatever may be the amount of bills or notes in the world, they add nothing whatever to its aggregate wealth, though they materially assist in its tranference from one man to another.

Apologizing for this little digression, I recur to my statement that it would be essential that part of the capital of every emigrant should consist of coin, and I would assign 5*l.* in every hundred to this object as an advantageous investment. This would probably be much more than would be required for the circulation of their property among themselves in the earlier stages of their career. But whatever excess there might be at first, would cease to be a superfluity so soon as they had made any considerable increase of their property, as we shall afterwards see.

It must be remembered that money being simply a commodity like everything else with which it is exchangeable, is limited in its uses, just as clothes, chairs, tables, or spectacles are. If there are more of them than individuals or the public require, the overplus becomes useless. They remain, in fact, incumbrances in the hands of the proprietor; or if he seek to get rid of them, he must do so at a much reduced price.

Now the use of money, though not superseding the principle of barter, most essentially assists it, by supplying that convenience in which barter is deficient. It is the authorized and universally recognized standard by which the value of different commodities may be fixed. If a man wishes to exchange a horse with another for a cow, and there is a difference in the respective values of the two things, something must be given by the owner of the less valuable article to the one who possesses the other. Without money this would be very difficult. The one might not have the particular goods that the other desired to have; and if he had, there must be a specific value put upon them; and even then, an accurate proportion of the difference in the values of the objects of the dealing, might not be satisfactorily arranged. But by a resort to money, with its capabilities of the minutest subdivision, the transaction might be at once completed. Money bears to traffic precisely the same relation that an interpreter bears to two foreigners who desire to carry out a bargain. The interpreter has nothing to do with the terms of the contract; he merely assists in its arrangement and completion. So, the use of money is simply for the purpose of carrying out bargains between buyers and sellers of all property, whether lands, goods, or services. It is the great balancer, as it were, of accounts between individuals, without in the slightest degree varying the figures in those accounts.

It is important to remember that this is its sole use; and if more of it exists in a country than is necessary for this purpose, it ceases to be of the same value; it becomes like other redundant things, cheaper, or what is the same thing, fewer goods can be obtained for any particular portion of it, than before the superfluity existed.

Nevertheless, coin, unlike bank-notes, forms a part of the substantial property of the country. It must be so, because the metals of which it is composed, have cost a considerable amount of labour to procure them, and they themselves are of use and service for many other purposes. Their value, in fact, as metals, very nearly approximates to what it is before they are coined; but it is necessary that the coin should be worth more; for if, otherwise, money were to become cheap in consequence of a sudden redundancy—and this not unfrequently happens—there would be a temptation to melt it down by those who could make more of the metal than they could of the coin. Besides, the labour expended upon it in the process of coining would always give it an augmented value beyond that of the rough metal.

But although coin forms a part of a country's wealth, it represents but a very small fractional proportion of the bulk of it. A very little coin is made to go a very long way. In small daily transactions it is indispensable, as we all know. Where there are larger ones, mere differences in value are adjusted by means of transfer to the seller of debts due from these parties to the buyer; and here we see how bills and notes supersede the necessity of using hard money. If A. owes 100*l.* to B., B. owes the same sum to C., and C. is similarly indebted to A., the debts of all may be easily cancelled without a single coin passing. A man may pay away a sovereign in the morning, and in the course of the day it may pass through the hands of twenty different persons, each receiving and parting with it in exchange for some equivalent in the shape of goods or services. It would thus have done duty for an amount twenty times its own value. A man in receipt of 100,000*l.* a year, seldom has perhaps more than 20*l.* worth of coin in his individual possession at any one time. He may obtain it from his agents or his banker whenever he absolutely requires it, if he ever does: while these again never have in their custody nearly enough coin to satisfy all their liabilities, but they know where to get it in their turn, from those who have a stock of ready money by them. At all events the 100,000*l.* has been under the control of its owner for a year. To render it profitable and get it replaced, he must have probably paid away and received back in some shape or another the whole of it, as I have before endeavoured to show; and yet a very infinitesimal sum in comparison with the gross income is sufficient, as far as he is personally concerned, to put the latter into circulation. All the gold and silver in the world might fall infinitely short of satisfying the debts due from one single country to the various other countries of the globe.

Transactions to the amount of many millions of pounds annually, are settled, and transfers of property made, with scarcely a necessity for the intervention of coin.

I have been thus explicit on the subject of money, because, however needless may be the explanation to many, there are always some who are disposed to attach much greater importance to it, specifically as such, than it deserves.

I have, then, stipulated for a little company of

twenty men, possessed of a capital among them of 2000*l.*, consisting of all that may be required to maintain them for the present, with ample means of augmenting their stores for the future. I want it to be understood, in fact, that they have possessed themselves of all the elements that would suffice to develop to the utmost, the resources of the soil on which they are located, and to assist in producing whatever it is capable of bringing forth. I assume, in fact, that their working equipment is complete. I must propose, too, that each man of them is, in his pecuniary circumstances, entirely independent of the rest, striving to do the best he can for himself, so that there may be constant emulation and competition amongst them—for it is these that constitute the strongest stimulus to exertion and industry. Of course they would soon arrange some simple form of government for themselves, to prevent any encroachment by some, upon the rights of others. They would probably lay down certain rules and regulations which, when once sanctioned by general consent, would have the effect of laws, and which the whole body would be pledged to maintain; so that if any individual were disposed to transgress them, the other nineteen would combine their social and physical strength to coerce him into obedience; and this they would be bound to do, so long as the law remained unaltered, though some might think its principle inexpedient or unjust. This is in fact, the fundamental theory of all well-constituted governments, although in these very recent and enlightened times it is not always acted upon. However, my object at present is to discuss economical subjects and not political ones.

Having thus given my *protégés* a fair start in their new life, I will try to suggest the course they would probably adopt in their pursuit of wealth and prosperity.

They would first construct huts, just sufficient to give them shelter from the inclemency of the weather, but with the appliances with which I presume they have provided themselves, this would be speedily accomplished. Their next consideration would be how they could most effectually provide food and other absolute necessaries, by the time their present stock was exhausted, as in course of time it must needs be. They could afford to run no risk in ensuring such a result, and therefore they would probably all be mainly employed in working to this end, by preparing and tilling the land, sowing it with seed of various kinds, tending their cattle, and generally doing what was essential for propagation and increase of their live stock.[1] This would occupy all their efforts for a certain period, and they would have little leisure for anything else; but as the stubborn earth yielded by culture more copiously to their exertions, as it certainly would do, it might happen that one of them, more skilful or more industrious than the rest, would find that he could raise twice as much produce as was sufficient for his own immediate consumption; and if the whole of the company were fairly provided for in the way of food and other necessaries, it is clear that his superfluity would be useless to him, unless he could in some way dispose of it when obtained. He might, then, bargain with one of the

---

[1] I cannot better illustrate the difference between organic and inorganic matter than by supposing that one of the emigrants had brought with him a small quantity of silkworms' eggs, which, to a casual observer, would appear little else but mere specks of dark sand. But mark the difference. The grain of sand would remain perhaps the same to all eternity. It could produce nothing. It would neither grow in bulk nor weight, whereas the little (to all appearance equally insignificant) egg might, by virtue of the wonderful organization and natural machinery within it, under favourable circumstances and proper care, be capable of producing, eventually, sufficient silk and satin to clothe all the inhabitants of our little colony. But of course no such employment of time and labour would be thought of until a much later period of their career, when the cultivation of luxuries had become a necessity.

twenty who had more capacity for carpentering and building than for raising food, to build him a house, and in return he would supply him with subsistence while he was so occupied. It would suit the purpose of both, for the one would utilize the extra food which he did not want, and the other would be engaged in a labour much better suited to his ability. At the end of the transaction, then, the employer would have got a substantial house for the surplus food which was perishable, and the other would be probably much better fed than if he had toiled to procure food for himself.

Now, if in the meantime the rest of the society had managed to replace in substance, all that they had consumed, it is obvious that there would be in the community the value of this permanent structure added to the total property with which they had commenced their enterprise.

But if one man could produce so much more of the actual necessaries of life than he could possibly consume, others might be soon enabled to do the same, and it might in the end be found that five men would be sufficient to raise an amount of produce that would maintain the whole twenty. The result would be that the remaining fifteen would be released from the necessity of toiling for their own food, and could devote themselves to other occupations—that is to say, to the manufacture of articles that would be useful and serviceable, as distinguished from essential, to the general community. They *must* apply themselves to some species of toil, for the food producers would require to be paid in some way for the aliment they had produced in excess of what they wanted for themselves. It would be by this process that the society would become enriched.

There could be no reason why their original stock should decrease. That part only would be consumed and lost to them, which by the hypothesis might be replaced by the co-operation of the energy and industry of man with the bounties of nature; whilst, on the other hand, there must be a constantly increasing accumulation of the various articles the large majority of workers had manufactured.

Of course there would be a constant interchange of commodities amongst the different inhabitants, so that all would be benefited by the possession of what was useful to them.

Now, if this process went on, it is easy to imagine that a time would come, when the whole twenty would be surrounded by all those objects of real service and utility that might contribute to their reasonable wants. They might all have commodious habitations, useful furniture, and so forth, and if every one had enough of things of a serviceable description it would be a mere waste of time and labour to manufacture more. The builder of houses, the maker of furniture, the man who made jugs and basins—each would find his occupation gone, for there would be no one to take them in exchange, where every one was already furnished with an ample supply. Still, they must work in order to produce something which they may exchange for food—since the agriculturist would take care not to grow more than he wanted for his own use, except with the expectation of getting something in return.

Now this brings us to the subject of luxuries, and we shall see that these are absolutely forced upon the society by the very exigencies of its condition.

But before I enter upon this topic, I should like to say a word as to the general distribution of the increased wealth that has accrued to them thus far.

It is only fair to presume that each ought to share in the general prosperity, and if all did not do so equally, they certainly would do it substantially, for all must now be taken to be fairly industrious, and all have been employed in producing only such things as were necessary or useful.

They might have gone on the principle of every

man producing everything he required for himself, and then it may be said that as a quarter of the time and labour of the community might prepare enough food for the whole, so each, if he were to procure it for himself, might do so in little more than a quarter of the hours he could give to labour; thus the remaining three-quarters of his time might be employed by him in making what would hereafter conduce to his comfort and convenience. He would then be precisely in the condition of Robinson Crusoe on his desert island, where he had to be his own provider in every species of requirement.

But the society would soon find out the advantages of the division of labour, and the additional facilities that practice and experience give to those who devote themselves to that particular class of work for which they are more specially adapted. Each would find it to his interest to confine himself to one species of production, and then to exchange his goods with his companions for what they, acting on the same principle, could furnish him with, and as, on my present assumption, nothing was made but what was of real utility, all would desire to possess them, and so every man would exchange with the others until the wants and requirements of all were satisfied. Thus there would be a full participation in the joint profits that the society had accumulated, and such articles of convenience that each possessed would represent his share of the total proceeds.

Now I have endeavoured to show that at the outset, all would probably be compelled to employ themselves in seeking the actual necessaries of life, but that to this production there would be a definite limit; for to extend it beyond what could be consumed, would be mere waste. This would be the first stage in their career.

The second represents them as having secured for themselves a thorough immunity from the risk of famine, with leisure to devote themselves to the acquisition of those articles of comfort and convenience that would render their labour more easy, and their hours of relaxation more pleasurable. But there would also be a distinct limit to the production of these things in the sense in which I employ the term—just as more chairs in a house than were ever likely to be sat upon, or more fire-irons than there were fireplaces to put them in, would be incumbrances rather than objects of utility.

If, then, every one had a sufficiency of this species of property, the demand for labour in that direction must necessarily cease, and it would become requisite to introduce some new occupation, some fresh incentive to the energy and industry of the workers, with a view to gratify other desires dictinct from the acquisition of what was merely useful.

We are thus landed in the third stage of our society's career, which, I repeat, renders the introduction of luxuries inevitable, unless the members are to become idle and their lives and faculties profitless.

No doubt it is very difficult to draw a precise line between what are mere comforts and conveniences and what are luxuries, and this must always be so, where contrary extremes gradually approach each other, until they glide imperceptibly into some mean or vanishing-point, equally distant from that from which they both started. For instance, it is not easy to detect the precise medium between bright sunlight and total darkness, but we can all appreciate the difference between day and night. We cannot fix a clear definition of where wealth ends and poverty begins, but we cannot be far wrong in stating that a man with 20,000*l.* a year is rich, and that one who has to support a large family on ten shillings a week is poor.

We might fairly contend that sleeping on a feather-bed would afford more comfort to the sleeper than if he sought repose on bare deal boards. But if he had the advantage of reposing on the former it would not be in the slightest degree enhanced by the fact that the bedstead was richly carved, or the coverlet an elaborately embroidered one. These more costly attributes would not increase the chance of slumber, or give the least accession of bodily enjoyment to the recumbent individual. Now what I have just said may perhaps afford as clear an illustration as could be given of the distinction I am anxious to draw between the necessities, the conveniences and comforts, and the luxuries of life. A sheltered spot, on which our bodies may recline, though it be a deal board, may be considered a necessity—a feather-bed beneath a snug domestic roof would be a convenience and a comfort, any expensive and decorative accessories, such as I have alluded to, would be luxuries. It is certain that we are endowed by nature with the faculty of appreciating and delighting in beauty, in taste, and refinement, with regard to things as well as to persons. But it is not given to all men alike, because such tendencies form rather the ornaments than the utilities of life, and as we have seen, what is essential to us has been bestowed much more liberally than that which is merely superfluous, though it may be subservient to our general welfare. The sense of what is useful comes home to every man. It assists him more advantageously to pursue his career, whatever that career may be.

We do not all perceive the excellence of a picture by Raphael, or a composition by Handel, but we all see, at once, how useful is a chair, a knife, or a timepiece. We might exist in the world quite as well for all practical purposes without pictures or melodies as with them, but we should find it very inconvenient to be without the others. No prudent or rational man would, I presume, store his house with expensive paintings, statuary, or plate, until he had possessed himself of all the substantial comforts and conveniences of domestic life, but when he had done that it would be folly to multiply them. A man carrying one watch about him might ascertain the time just as well as if he carried half a dozen.

But when the householder was fully equipped with everything that might satisfy his real wants, he would be quite justified, if he could afford it, in indulging himself with luxuries. A watch would be of use; a picture, or a piece of sculpture, would only serve to gratify a fancy; but still he could spend his surplus income in no other satisfactory way, if he were resolved to spend it at all. He might (if a member of a commercial community) invest it in trade; or, if he did not choose to do so himself, he might enable others to do it, by lending the money out at interest; and it is pretty obvious that the borrower, who paid him that interest, must in some way or another have so employed and utilized the loan, as to make it produce much more than the annual sum he paid for its use —or he would gain nothing by the transaction.

Of course there are many different species of luxuries. They are the mere indulgences of a taste, a fancy, or a fashion, and to these we know there is no limit, such as we have proved to exist with reference to necessaries and mere conveniences. They vary with the individual, and even with the times. Some of them trench, as I have said, on the other two elements just mentioned. Tea, as an instance, comes very near to the boundary-line. At its first introduction, it was regarded essentially as a luxury, accessible only to the few. Now, the poorest family in the kingdom would deem itself poorer, if it were to be deprived of that particular beverage. In fact, there are few households, even the most lowly, where what might be fairly termed luxuries—in the sense of things that might very well be done without—have not by habit and custom become matters of daily use. If tea is an example in one direction, wine may be quoted on the other hand. It is tasted but seldom by a large mass of the population; and some very wise people are always trying to persuade us that if the world were to get rid of it altogether, it would sustain no great loss. At all events the broad distinction between these two species of property, the useful and the luxurious, is tolerably clear.

I left our adventurers at the fulfilment of their second stage, when an appeal to luxuries was about to ensue, and I hope I have proved that a resort to them is now actually essential to a continuous and advancing prosperity.

With industry and energies still unflagging and urging them onwards to some new objects and desires, for they have satisfied their old and more pressing ones, no other course would be open to them than to seek to produce that which is pleasing, instead of what was heretofore simply useful.

They had been accustomed to the habits and tastes of civilized life before they embarked on their novel expedition, and would now strive after ideal and sensuous gratifications, after having achieved the more real and substantial ones. They would probably in the first instance betake themselves to the task of ornamenting and rendering more elegant and tasteful the objects of utility, of which they now had an abundance: plain wooden chairs and tables would become chiselled and polished one; clothes would be made of more expensive material and of more studied form; eating and drinking vessels would assume a more artistic shape: men would rack their brains in devising something that might be attractive to their neighbours, and each would strive to do what, by gratifying the wishes of the rest, would, by the process of interchange, obtain that which would be pleasurable to him; and, I repeat, if they were still to work and toil, they could do it in no other way. It would be the only mode by which their industry could be made available in promoting their own enjoyment. For instance, one might have a talent for painting, another for sculpture, a third for music, &c., a fourth for house decoration. By constantly bartering, or buying, or selling the results of their several capabilities, one with another, all would derive gratification from the labour of each. And, roughly, it may be said that the price put upon these various productions of skill and taste would be reckoned by what was consumed by the makers in their production. It must not be supposed that I mean by this, the mere necessaries that supported the artist, whilst his picture was being actually painted. A painter or a sculptor must have spent a considerable time in learning his art : during this period his labour would be entirely unproductive. When he was able therefore to produce anything worthy of his craft, he must needs be recouped the expense he had been put to while acquiring his skill.

As far as the total funds of the community were concerned, the learning to paint, as well as the painting the picture, must have absorbed and destroyed a large portion of them. But what was so consumed, was the property of the artist himself, and must be repaid to him by those who purchased his work; and the picture would be all the society, at least, would have to show for the large amount of sustenance that had been virtually expended upon its execution. Though the painter got the price, the purchaser would lose it; and whether that price was 5l., 10l., or 100l., the aggregate funds of the community would be unaffected by the sale, whilst they must have been considerably diminished by the process by which the object of it was procured. But I assume, of course, that the colony had arrived at that stage of superfluous wealth, if

I may so express myself, that it could very well afford the expenditure. For, remember, that through however many hands the picture passed, and whatever might be the successive amounts given for it, it would be simply the exchange of one thing for another, each remaining the same.

To illustrate this matter a little further, let us think for a moment what would be the effect of one of the inhabitants finding on his land a jewel of such size and quality that it would fetch a very large price in the old country. It would certainly form no substantial accretion to the total wealth of the new one. As long as the society remained isolated, it would be something like a picture, a mere additional possession, but in truth a much less important one. It would merely afford a transient pleasure to the sense of sight. But pictures, while they do this, would in some measure give instruction, and even improvement to the minds of many persons, by calling into action other faculties that were intended to be exercised. They might, for instance illustrate history, and help to fix its events upon the memory; they have served to stimulate men to deeds of valour and patriotism; they may awake emotions that tend to soften the heart, and lead to repentance and virtue. In this sense they might be regarded as things of use and convenience, if we were discussing mere moral topics, instead of economic and pecuniary ones. But a jewel (and I of course treat it now exclusively as an ornament) could only serve to gratify the pride or vanity of its possessor, although even these qualities are not to be treated with scorn, since they are highly valuable when kept within proper bounds. We see around us every day things equally beautiful with rubies and diamonds, and which are much more useful, but they are not so rare; and people may always be found who will pay large prices for any rarity, whatever may be its nature. An individual will give perhaps 1000*l.* for what he is assured is a fine brilliant, though he might be unable to distinguish it from a cleverly executed model made of common crystal. It is evidently a sentiment—a mere fancy that allures him.

Now to have found the jewel I have spoken of, might be a lucky incident for the finder, because there might be some of his companions who would be disposed to give him a large price for it, either in money or goods; but this, again, would be a mere exchange, involving no general increase; what he would get, the purchaser would part with; and this would be the case, however many buyers and sellers there were. Our community, remember, is a very limited one, and therefore the opportunities for trafficking are limited too. But the world also may be said to have its limits in this respect; and no amount of precious stones or jewellery would augment its totally aggregate wealth.

If our adventurers had some separate body of people near them with whom they could trade, some of them might be willing to give many articles of use or luxury to purchase the jewel. Our society as well as the finder might be all the richer by the exchange, since the articles coming in as payment might be very available for further production, and must at all events save the labour that, if the colonists had to make them for themselves, they would be obliged to expend. But in proportion as one community became richer, the other would be poorer, and taking them together there would be no more wealth amongst them than there was before.

The jewel might wander round the world, constantly changing owners; the price that was paid for it might be increasing on every sale; but the same principle would apply: what the buyer gave the seller would part with, and the jewel at the end of its travels would continue precisely the same as ever, a mere thing of beauty but of no real utility. The property that had been given in exchange for it would be differently distributed to what it was before, but it would not be augmented.

It may be truly said, then, that a knife, a saw, or a spade would be of infinitely more service to the community than the finest emerald or brilliant that was ever found within the bowels of the earth. They might daily and hourly for years be used in their humble way in profitable operations that it would be difficult or impossible to carry on without them, while the jewel would remain as a mere embellishment to be worn and admired.

It is probably some such comparison that has engendered the notion before referred to among a certain unthinking portion of the labouring classes, that they are much more beneficial to society than the rich man who lives at ease upon his income. The latter they assimilate to the jewel, and themselves to the tool—they do all the work, and the other does nothing towards production. But they forget, in the first place, that the tool is valueless of itself, and can only be made available by having some material in improving which, it may be employed; and in the second, some skilled agent who can wield it to advantage. This material as well as the agent requires pre-existent money, for the one has its price and the workman requires his wages while he works, and without these he might starve.

Now it is capital—stored-up labour—that can alone supply these essentials, and it is the rich man that possesses it. Personally he may be insignificant, weak, and worthless; but he has the wealth that can set industry and skill in motion. Without him the implements and the workmen would be idle, useless, and without occupation; and I trust I may take it as proved already, that in whatever way he uses his wealth, whether prudently or recklessly, he can scarcely help employing it in the support and maintenance of a vast number of persons, who make substantially as much profit out of it, for their purposes, as he makes for his.

Still, as I have said, the influence of this peculiar class of luxuries, such as jewels, on the world's action, is not to be overlooked or despised. The desire to possess them has been productive of much good as well as of much evil. They have induced many to labour in various ways—commercially, scientifically, and politically—seeking only perhaps to aggrandize themselves, but incidentally contributing largely to the benefit of mankind. On the other hand, they have led to much crime; and in the history of one celebrated diamond we learn that the wish to possess it was the motive to the murder of more than one prince or rajah.

The desire for wealth may be like most other desires; by excessive indulgence it becomes a vice, but kept within proper and legitimate bounds, by rousing men's energies and urging them on to labour, it carries out one of the most useful and beneficent laws of nature.

What I want strongly to inculcate is that a temperate indulgence in luxuries is not the mere gratification of a whim, or a wasteful expenditure of the bounties of Providence. It is a necessity forced upon us by the very exigencies of our being, and one which, while it affords pleasure to the rich, is of immeasurable advantage to the poor. The very disposition to possess luxuries begets the fund that enables their possessor to enjoy them, and that very fund maintains and nourishes the labourer while working to produce them. At all events, we have seen that our colony must enter upon a career of luxury, or it must stagnate. It would virtually soon return to its primitive condition of each man providing for himself that which would satisfy his immediate needs; all buying and selling would speedily be at an end, for while all were supplied with as much as they could wish for of necessaries and conveniences, it could scarcely be otherwise. Three-fourths of their time would be wasted in indolence and inaction. The food producers would cease to produce more than sufficed for their own consumption, since they could not part with any surplus, except for articles of which they already had an abundance, and every other man, whatsoever might be his occupation, would be precisely in the same position.

I need not dwell on the various species of luxury that might excite desire in some, while it stimulated others to labour. Many things would remain durable and permanent after their production, such as pictures, carvings, or other works of art. They would be valueless so far as any accession to the aggregate wealth of the community was concerned, as long as there was no external market in which they might be exchanged, although, should this opportunity ever arise, they might fetch ten times the value of the material and time consumed in their formation, and thus the property of the society might be augmented.

Many things might be produced which merely addressed themselves to the appetite, and which would be destroyed in the very fruition of the enjoyment they were meant to afford : such as wines, hot-house fruits out of season, which would require a large outlay for their growth. These would be consumed, and leave no trace of their having existed. One can readily see that the production of those luxuries which were permanent might be more advantageously manufactured than those which were thus evanescent. But if the society had reached that point to which I have sought to bring it, where every individual was surrounded by sufficient articles of convenience and comfort, I repeat, that its members might soon be able to indulge in transient gratifications, the producing of which had at least given employment for a time to others, and prevented them from leading a life of inactivity and sloth.

Again, if the principles I have enunciated are sound, it is obvious that the property which the society possessed at the outset would in the course of time be vastly increased, and the increase would be due to the surplus time that would be left to them in which to manufacture lasting and substantial things after they had provided themselves with what was necessary for their actual subsistence.

But it may be worth while to state here that in their own estimate of the money value of their possessions it would amount to a denominational sum, very much less than it would be valued at, according to the currency of the outer world. They could only reckon its worth according to the exchangeable value of the money that circulated amongst them. If that value was altered, the nominal value of their property would be altered too, and we have seen that the current price of money varies with circumstances. If there is a scarcity of it in comparison with other things, its value rises, and *vice versâ*. I have assumed their original property to be worth 2000*l*., and that there was amongst them 100*l*. in coin, and I will take it for granted that they could not get more. Now although that amount of specie might suffice to circulate a property worth 2000*l*., it would probably be insufficient to sustain the burden of circulating a capital that had multiplied tenfold. The property had increased; the coin had remained stationary. There would be a much larger demand for the latter than the supply could meet; money would become dearer, that is to say, a greater quantity of goods could be obtained for any particular coin than would have been obtained before.

But the same principle would apply to the whole property as to a part of it, and its increase in nominal value in money, would be very much less in proportion, than its real one. So that if their possessions were ten times greater than they were at the outset, they would be probably reckoned by the settlers themselves at not much more than their original price. We are so much in the habit of attributing dearness or cheapness to the articles purchased, without regard to any change in the value of the coin paid for them, that we are apt to lose sight of the real nature of any variation in prices. Dearness of commodities, for instance, may arise from one of two causes: a scarcity of the commodities themselves, or a fall in the value of money. But where it proceeds from the latter cause, the increase in the price of goods would of course be general.

If, on the first day of any year you can buy a sheep for 1*l*., and—the demand and supply remaining the same—if, on the first day of the next year you have to pay 2*l*. for one, it is fair to assume that the difference is due to a change in the value of money, and that it is by reason of a large influx of bullion, or perhaps other causes, that it has become less valuable; or, in other words, less will be given for it than was given twelve months ago.

In the case of our colony, while the objects of traffic among themselves were largely augmented, the means of conducting that traffic through the medium of hard cash, had become diminished. There would be more transactions in the way of buying and selling, paying for services rendered, carrying out bargains, and so forth; and whilst the conveniences for doing these things had considerably decreased by the scarcity of money, what there was, must needs rise in exchangeable value.

I hope I have explained how in estimating the value of their property by the only standard they could adopt (the nominal value of the money that existed among them) they would reckon it at much less than it was intrinsically worth. It is only important as further illustrating the principle that money is subject to the same economical laws as other kinds of commodities, and that its exchangeable value depends upon the supply relatively to the demand.

Whenever the colony again came into commercial relations with the world at large, the ratable value of each coin would sink to its normal condition, property would rise in price, and the currency would fall in value, in accordance with general prices established elsewhere.

In treating of the progress of our emigrants towards prosperity and wealth I have assumed them to be uniformly skilful and industrious, each seeking to aggrandize himself, though by that very means necessarily contributing to the welfare of the whole body. My object was to show what it was possible to do, rather than what it was likely would be done.

But it is pretty obvious that although they all began with the same amount of capital, inequalities in the distribution of wealth would soon arise amongst them. Some would be more assiduous, more energetic, more enterprising than others, and might therefore become more prosperous and comparatively wealthy. Some might be idle and unthrifty; and some again, without any fault of their own, might by accident or misfortune be reduced to poverty and privation. The cessation from labour, however, caused on the part of one or more individuals would of course prejudicially affect the gross accumulation of the funds possessed by the whole body; for instead of adding to those funds by the produce of their toil, the idle or unfortunate would be consuming, and so diminishing them. They might have a moral claim on their more prosperous brethren, but no one could affirm that the latter had not an absolute right to that which their toil and industry had procured them.

A sort of difference of rank, too, would soon be established; for inasmuch as wealth is power—at least, in the sense of enabling its possessor to confer benefits on others—that power would always beget an outward respect and consideration, and that without necessitating any sacrifice of independence on the part of those who exhibited it. Where homage is not insolently exacted as a right, it will in general be cheerfully manifested where it is due; and when a man has the means, the opportunity, and the will to oblige others, it is not unnatural to suppose that he would be looked up to with extra consideration.

The aristocracy of wealth is generally regarded as the most despicable of all social dominations; but this arises probably from the fact that it is not infrequently accompanied by sordid views, upstart manners, and impertinent airs of superiority. Yet this is by no means an essential or invariable attribute of the possessor of riches, and discuss the matter with as much philosophy and as much affectation of manly independence as we may—as long as human nature is constituted as it is, large possessions will always have their influence; and perhaps it is well they should, for nothing can afford a stronger guarantee that the owner is an advocate of order and good government and a staunch supporter of a sound and well-devised constitution.

But I only desire to suggest here, that in all

probability there would soon be a difference in grade amongst the inhabitants of our island. One important change in the distribution of their visible property would certainly soon arise, and that is with regard to the possession of land. I have supposed that, in the first instance, there would be a tolerably equal partition of the soil among the whole body. But it would in a very short time be found that one species of occupation required much more of the surface of the ground for its operations than another. Some would require merely enough to contain a house or a workshop. Those who turned corn growers or graziers would require much more than they obtained from their original allotments. Those who had too much would part with the surplus—of course, for a consideration—to those who had too little, and a share that once belonged to a single individual might be cut up into small parcels, each the property of a different owner; but it is clear that whoever was substituted for the original allottee would stand in his place, and have, by the purchase, all the rights he once possessed. Now let us inquire how it was that any one of them ever gained a paramount, indefeasible right to a portion of the soil. He could scarcely base his title on the mere fact of having been the first to take possession, or that, by the general assent of the community, his particular part of it was allotted to him.

Suppose a second colony had come upon the island the day after the first had arrived and had taken possession. It might perhaps be difficult to contend that the first comers had an absolute right to exclude the others from a claim to the smallest fraction of the soil. Mere possession for twenty-four hours would seem hardly sufficient to confer on them an exclusive title against all the world. But so soon as they had expended capital or labour on the land, they would be in a very different position. They would then have a clear advantage over strangers by every principle of justice and equity.

Suppose they had improved it by draining, manuring, planting, by clearing the ground, and building huts or houses. After all this had been done, it would surely shut out fresh invaders from all claim to participate in benefits to which they had in no way contributed. The improvements would be inseparable from the soil on which they had been made, and without the most ample compensation for previous outlay, it would be to the last degree unjust that the improvers should be disturbed in their occupation.

Now, it is virtually on this principle that the title to all lands in the possession and ownership of individuals is based. Either they themselves have reclaimed and improved the soil, or their predecessors have done so, and they have legitimately succeeded by purchase or otherwise to their rights, and what had been paid is merely equivalent to that which has been heretofore expended, representing the hoarded accumulated labour of the purchasers, as has been before explained.

Ideas couched in vague generalities of expression are often promulgated by those who hope to profit by the theory, that every man born upon the soil comes into the world with some vested and inherent right to a portion of it.

If what these people mean were subjected to the trammels of a definition, the proposition would at once refute itself. If every one had a right to a part of the soil, it is obvious that no man could have a continuous title to any particular fixed portion of it, since a change of population must be constantly necessitating a redivision.

The whole community is interested in having the land cultivated and improved; but in doing this it is often essential to bestow upon it large expenditure for which there can be no return for a considerable time—often for years, as in the planting of young trees for timber.

Who would invest his capital in that of which he might be dispossessed long before any profit from it could accrue to him? Without a guarantee that we should be secured in the enjoyment of the profitable results of our toil, we should be reduced to the primitive condition I have before described, where men would live from hand to mouth, content to produce what was merely necessary for their subsistence, since any property that they saved or accumulated might be filched from them by the hands of strangers.

The allegation is that the land is common, and that any person condescending to be born on it, acquires an irrefragable claim to a portion of the surface. Let us admit this for a moment to be true. But at all events, if it is once conceded that there can be an absolute right to anything, the produce of a man's labour must be held to be his own. Fish in the sea are in general the property of any one who will take the trouble of catching them. They may be said to be common property in the first instance, but the man who devotes his labour to reducing them into possession has an incontestible right to call them his, against all those who before had as much interest in them as he had. Even if land were originally the property of each and all, still where capital has been irrevocably bestowed upon it, and has rendered it more valuable —and that, in accordance with a law of nature, that it was designed for cultivation—it must lose its universality of ownership. It is said there is a land hunger as there is a food hunger and a money hunger. I might go further and say there is a watch and chain hunger on the part of many of our fellow-subjects who prowl at night about the dark purlieus of the metropolis. But if this abnormal craving cannot be appeased except at the expense of every principle of justice and common sense, it must be forcibly checked and restrained, or at least, no additional facilities should be afforded for its gratification.

PART II.

I HAVE dealt with our emigrants hitherto, as if they were entirely separated from their fellow-beings in any part of the world. I now propose to treat them in connexion and correspondence with other bodies of people with whom they have free intercourse for purposes of trade and commerce.

I will suppose, therefore, that they are now surrounded by islands with communities somewhat similar to their own, except that there shall be varieties of soil and climate, and that in consequence each shall have some peculiar facilities for producing certain commodities, whether natural or artificial, that the others do not possess; and I must also premise that the several other islands had been subject to no restrictions as to intercourse with one another.

Now, whether the change were to come suddenly upon our old acquaintance, or that there was no change at all, but a mere continuance of what had long existed, is not very material, inasmuch as in the first case, the principles of interchangeability would speedily assimilate, and the prices and value of property would be pretty generally equalized throughout the various populations.

No doubt, if it were sudden, the change at first would be very great. Our original society would find its accumulated property very much increased in nominal value, for the restriction in amount of their coin, which must have hitherto been their standard of price, would be withdrawn. The scarcity which had given it a fictitious value would cease; for at every sale of any of their possessions to their new neighbours coin at the old world estimate would flow in, and soon reduce theirs to its normal condition. No outside customer would take their money at the high rate they had been compelled to put upon it, and no one of their former brethren but would sell his goods to the foreigner in preference to his neighbour, since he would necessarily receive from the former a larger sum in specie. The holder of cash would be impoverished by reason of the large accession of coin, and the consequent decrease in value of what he held at the time. The proprietor of goods would believe himself enriched, because he would get much more for them nominally than he could previously obtain; but he would find himself mistaken in the end, for the extra sum he received would go no further in purchasing, than the smaller one he could sell at previously.

But what I have just said has reference only to the change in the value of money. The transition from a state of solitary existence to one of general association with competitors in commerce might render the inhabitants in effect much richer, substantially, than they were previously. Their accumulated productions might exhibit such skill and excellence as to excite the admiration and desire of their new customers. As I have before hinted, the latter might be so charmed with a picture or a piece of sculpture that they would give them twenty times the real value of the labour it had cost to produce it, and this cost, as we have seen, would in general be the quantity of food and other necessaries that had been consumed during the whole process of its formation, including, of course, a proportionate share for the education of the artist. This would be without regard to the nominal price of money. Whether the cost of the food, &c., were more or less would be immaterial. It would be the quantity disposed of that would fix the real value of the article.

Now, if the acquisition of twenty times the value of the necessaries used up, and so destroyed in the course of its production, were to accrue to the painter or the sculptor, it is obvious that he would be much more wealthy than he was before. Other results of skill and labour might be coveted in the same way, and be similarly paid for, and the proprietors would of course be largely benefited too. But what I wish to be particularly observed (though the remark is an obvious one) is this, that not only would there be a great gain to the individuals who had wrought these things, but the aggregate property of the entire community would be largely augmented also, and to promote this result is peculiarly the province of the science of political economy. If the consideration were given in money as, to the individual seller, it probably would be, the gain would be appreciable by every one. The real transfer as far as the two populations were concerned would be of commodities rather than of specie, and our colonists would be relieved from the expenditure of labour so far in producing them, and that labour might be then expended in distinct and profitable ways. This consideration will be seen to be highly material when we come to discuss the arguments that are often used with reference to the advantage of employing native labour instead of availing ourselves of that of strangers.

The wealth of a community must be made up of the separate wealth of individuals. If you increase the latter, you naturally augment the former; unless, indeed, when by monopolies or artificial imposts and restrictions the State seeks to enrich one class of men at the expense of the general body.

It would perhaps be universally conceded that the best mode of ensuring the wealth and commercial prosperity of the inhabitants of the various islands I have assumed to be in communication with one another, would be to allow them free and unfettered liberty to interchange their several productions, every individual buying as cheaply as he could from any other, regardless of whether he were a stranger or a neighbour.

Divergence of opinion only arises when any one society begins to place difficulties in the way of this free interchange, by putting restrictions in the shape of duties on the importation of foreign productions into its own territories. The motive alleged is a desire to encourage its own manufactures, and thus give employment to its own people. When this course is once taken, then arises the question on the part of the proscribed communities, how this injury to themselves (for no one denies that it is one) can be most effectually obviated or checked.

It is pretty clear that if a man can purchase an article for half the price at which he can make it for himself, and he persists in being his own artificer, the one half of his labour would be thrown away. The difference between dearer and cheaper generally resolves itself into a question of time, since price depends upon the *quantity* of labour expended upon the production of the subject-matter; and if its price is doubled, it would in most cases imply that twice the quantity of necessaries have been used up and so destroyed in making it. And if the man in the case above stated were to manufacture the article himself, instead of buying it elsewhere, the superfluous food, &c., he absorbed during the operation might just as well be buried in the ocean.

It is true that he had been nourished and supported during the whole time, but as far at least as the half of that time was concerned, there would be nothing in the end to show for it. He would have been consuming and destroying, without producing any tangible thing. Had he bought the article, instead of fabricating it at double cost, he would retain in his hands the precise amount of money he had paid, namely the half of the larger outlay he would otherwise have made; his time would be at his disposal, to work in other ways for which he was better adapted, and which would leave some object of value as the result. It may be a trite illustration, but it is a conclusive one, that if a shoemaker were to employ a tailor to make his clothes, and the tailor were to furnish himself with shoes from a shoemaker, instead of each making those articles for himself, they would not only be better clad and better shod, but they would save a large amount of time, cost, and labour. They would in truth be both benefited by buying in the cheaper and better market. But if this would be the result in the case of two independent men, each looking to his own interest, and caring but little for that of his neighbour, the same principle must apply to two independent nations, each looking exclusively to its own welfare.

Still the question remains unsolved—how is a nation which interrupts this free interchange of traffic by the imposition of impost duties on foreign goods to be treated by those whose merchandise is thus sought to be excluded from it. Is it expedient freely to receive what it may send to you, though it persists in maintaining restrictions on the receipt of yours? I answer unhesitatingly in the affirmative, and I think the policy of such a course may be proved almost to demonstration.

Anything that tends to check the extension of commerce must necessarily be evil, and for this simple reason, that the commercial exchange of one thing for another obviously implies a gain to both parties engaged in the transaction; for if this were not so there would be no inducement on the side of one of them, at least, to engage in it. Again, if it increases the wealth of the individuals, it must also increase so far the wealth of the nation to which each belongs. Now the very imposition of import duties is meant to diminish the opportunities for the free interchange of commodities to which they are applied, and it is an acknowledgment by the nation enforcing them that they cannot produce them themselves as cheaply as they could purchase them elsewhere, for no one would think of paying a higher price to foreigners for goods of the same quality than he could purchase them for at home. Admitting then superior facility of production on the part of the country against which the impost is directed the imposing nation is willing, for some reason or another, to throw away the extra labour and cost of manufacturing for itself. Now, whatever that counterbalancing reason may be, at all events as far as either public or private wealth is concerned, it is wasting instead of augmenting it.

It is a mere sacrifice of capital, and far from encouraging the employment of its own labourers, it effects a wilful diminution of the fund by which labour is supported. It is true that more labour may be employed for the moment, but a large part of it is unproductive, and leaves no trace of anything remaining after that which maintained it has been consumed. It would be precisely the same as if you had employed and paid wages to twenty men, for doing what ten perhaps could have effectually accomplished in the same time. It would be no satisfactory explanation to say that the extra ten men have been supported by their wages during the whole period. They had been consumers, but they had produced nothing as the result of that consumption. They might as well have lived in idleness, and fed gratuitously upon your bounty.

For a nation then to put a tax upon articles of manufacture in order to enable its own countrymen to be employed in producing them, necessarily at a much larger cost, is not only a wasteful consumption of labour, but it compels every member of its population to pay a larger sum than he need pay for the articles themselves; for it must be remembered that price is regulated by the quantity of labour that is used in the production of goods. It diminishes too the number of commercial transactions, and the mutual profit attaching to them, since if a country refuses freely to receive foreign goods in exchange for its own, it must result that its exports would be less than if no restriction existed. It is pretty clear then, that self-imposed duties on foreign commodities, in order to promote home industry, must needs be injurious to any nation resorting to them; and this seems to suggest the course that should be taken by those countries whose goods are thus refused free admission to the taxing one. They must participate to some extent in the loss, because they would miss the profit they had hitherto made by the existence of a market for their commodities that was now closed to them. This they must submit to under any circumstances. But what advantage would they obtain by resorting to a similar course against the foreigner, that he had pursued towards them? That which they suffered from was a contraction of their trade. What they proposed to do would still further contract it, by adding a restriction on their own side to that which had already been imposed upon the other, so that there would be two contractions instead of one. If one tradesman were to refuse to buy goods of another merely because that other declined to buy goods of him, and he persevered in this, notwithstanding he could purchase what he required at a much cheaper rate and of a better quality than he could buy them elsewhere, everybody would pronounce him to be a very foolish person. But if the tailor, because the shoemaker would not have his clothes made by him, determined to make shoes for himself, a lunatic asylum would perhaps be a fitter receptacle for him than any other. Substitute two nations for the two individuals, and why should not the conclusion be the same.

Now I wish it to be clearly understood that I am only dealing with those import duties that a country levies upon foreign goods with the sole object of establishing and encouraging the manufacture of that species of things for itself. In fact, that is the only form in which the question is ever raised. Reciprocity, fair trade, protection, all virtually mean the same thing, namely, a preference for native over foreign industry. To treat other nations as they treat us would no doubt be but fair, if it were expedient, and to promote the welfare of our own workman would be both patriotic and meritorious. But the question is whether the way in which we set about it would not do the objects of our solicitude more harm than good.

The motive is praiseworthy; my contention is that the means proposed are fallacious and would utterly defeat themselves. This principle of retaliation might gratify our spleen and our revenge, but we should find it a very expensive indulgence.

After all, commercial patriotism in general stops short at the point of self-sacrifice. Every individual will look to his own interest first; when that is provided for he will much prefer assisting a fellow-countryman to helping a stranger, but in this instance there is no antagonism between his own welfare and that of his co-patriots. They would both profit by availing themselves of the extra facilities that Providence has given to different localities on the face of the globe for producing certain species of commodities, whether natural or artificial. If other countries decline to participate in those benefits, it is to their own loss; why should we voluntarily subject ourselves to the same privation.

I confine myself to these ostensibly protective duties, because there are many others that are imposed upon foreign imports for peculiar and specific reasons. But these are exceptional and anomalous, and I am only dealing with generalities. There are many courses that would be totally indefensible upon abstract principles, but which might be expedient or even necessary under a particular state of circumstances. Duties, for instance, continue to be levied mainly for the sake of revenue, such as on foreign wines, tea, tobacco. There is no scope for competition on our part as to these, and they are not, therefore, prohibitive, as those I am denouncing must be to a great extent, or they would not effect their intended purpose: when they produce to the State large amounts, are of long standing as a means of materially assisting the revenue, and have led us into complicated relations with other countries, with a view to their maintenance, however impolitic they may be in principle, it is not an easy matter to change the system. Some duties again may be prohibitive and yet salutary and essential, and it might be expedient to prohibit some importations entirely,—for instance, certain poisons, highly dangerous explosives, or foul and licentious literature,—although a traffic in them might produce much profit. But I pass by all these things for the reasons I have stated, and only deal with those which are newly sought to be imposed for the ostensible purpose of promoting home production.

While upon this subject, I do not wish to lose sight of an argument frequently brought forward by those who call themselves Fair Traders; namely, that the gain to the revenue by such imposts as I am censuring affords a material relief to general taxation, and as the expenses of the State must be provided for in some way, this, they say would be an advantageous mode of accomplishing that end. Now there are three principles among others (as will probably be admitted) that should be always kept in view in every species of taxation—first, that it should be fairly and equitably distributed so that every member of the community should pay a share proportionate to his wealth, because it is obvious that the larger are a man's possessions, the larger is the amount of protection he requires for them. Secondly, that the mode of collecting the tax should be as easy, simple, and inexpensive as possible, so as to increase as far as may be its net productiveness. Thirdly, it should studiously avoid putting any check upon the trading or manufacturing industry of the country.

Now, with regard to the first point, direct is evidently more equitable than indirect taxation. The income-tax is an instance of the first. The tax upon tea illustrates the latter.

We used to put a tax upon very many articles that constitute the bare necessaries and ordinary conveniences of life—upon those things which the poor would actually require for consumption as well as the rich. With regard to these, I have before sought to show, that the rich man individually requires and consumes very little in excess of the working man. For instance, he will not eat much more bread, consume much more sugar, drink much more tea, and might possibly smoke less tobacco, than one who possessed one-hundredth part of his income, and thus he would contribute indirectly, a mere infinitesimal part of the aggregate tax in proportion to his means.

A man with 100,000*l.* a year under the direct system, assuming it to be at 6*d.* in the pound, would pay a sum annually of 2500*l.* It is pretty clear that under no scheme of indirect taxation that could be devised, could he be compulsorily mulcted in one-twentieth part of that sum—I say compulsorily, because he might always escape paying taxes on expensive luxuries, by not consuming them. It is true he would have to pay the imposts on domestic articles for his whole household, but this would not much vary the proportion, and he would escape paying for those independent agents, labourers, workmen,

&c., whom I have heretofore represented as substantially dividing the benefits of his income with him, since they in their turn would have to pay taxes on what they consumed. Take, as a further and final illustration, the very familiar commonplace article, soap. How much oftener would a millionaire wash his hands in the course of a day than a poor though respectable cottager on his estate? If not more frequently, then, if there was a tax on soap, he would pay no more to the Exchequer personally than his humble tenant. But though the income tax is the most just in theory, it has its drawbacks in practice. We all know that it is extremely unpopular, except amongst those who are relieved from paying it. It is looked upon as far too inquisitorial, and, moreover, it is frequently evaded to a large extent by false returns. By the other and indirect mode of assessment, the true nature of the impost is concealed from men's view, and they cheerfully pay the price demanded for what they require, oblivious or regardless of the fact that part of that price is really forced from them by the necessities of government; and to keep the public in good humour, or to prevent its lapsing into a bad one, is by no means an insignificant or objectionable attribute of statecraft. Comparing then the advantages and disadvantages of the various systems of taxation respectively, it is perhaps not inexpedient that some compromise should be come to regarding them, and I do not wish it to be supposed that I am finding fault with the mixed course that exists at the present day, though I would object to any scheme that sought to extend the principle of indirect taxation.

But of all the methods of levying taxes for the purposes of revenue that could be devised, the most pernicious perhaps is that which seeks to lay imposts upon foreign manufactures, for the express purpose of preventing competition with our own. It transgresses all the three principles I have laid down. It is unequal and inequitable, for those only pay it, who require the particular articles that are so taxed. If it were confined to pure luxuries, which might easily be done without, it would be somewhat less objectionable, because those who insisted on having them, could scarcely complain of having to pay extensively for their gratification; but as the scheme is propounded, it includes within its scope all manufactures, whether useful or ornamental, with which we could possibly attempt a rivalry, however unsuccessful it might be. Then, it requires an elaborate and costly machinery to collect it, as well as to carry out a harassing supervision in general, over merchandise sent into this country, whether contraband or not, and thus the produce of the tax is materially reduced.

But the most formidable objection is, that it cramps and restricts trade and commerce, by not only limiting importation, but, as a consequence, diminishing our opportunities and means of export.

Thus, the scheme would be accompanied by all the ordinary disadvantages of indirect taxation, whilst it would present many other evils exclusively its own. The profit derived from it to the Exchequer would constitute a mere fraction of the loss sustained to the country by the sacrifice of the many interchanges of commodities and the gain attendant upon it, that it would entail.

I have dwelt on this subject with perhaps too much persistence, too much iteration, and a needless sameness of illustration; but my excuse must be, that an argument that may strike one mind, may not influence another, and that a redundance of proof is preferable to a deficiency which leaves the reasoning obscure. Yet I think I may further simplify the views I have sought to lay down, by treating them in connexion with the populations of the islands I have supposed to exist, and which I assume are now in trading communication with one another. Let me then distinguish the one whose origin and progress I have attempted to trace, by the letter A, and a neighbouring one by the letter B, and I will suppose that there is the ordinary commercial intercourse between them. Let me further assume that A could produce corn at half the price, or, in other words, at half the labour that it would cost B, and that B could grow cotton at the same proportionate advantage over A, were A to attempt the same thing. It could scarcely be doubted that if A grew corn to supply both islands, and B did the like with regard to cotton, and that they mutually exchanged their surplus produce, there would be, at the end of a given period, a great saving of labour and a large increase of wealth to both. Had each remained isolated from the other, and had each grown both articles for itself, it is obvious that A would have only half the cotton, and B only half the corn it might otherwise have been in possession of. Now suppose that A were determined to grow its own cotton, and prohibit the importation of that grown by B; if B refused to avail itself of A's supply of corn, because A refused to take B's cotton, it would be precisely in the disadvantageous condition I have just described. It would be wasting its substance by growing corn under great difficulties, when it could easily procure it without encountering any difficulty at all.

It must be remembered that a nation, however anxious to exclude foreign manufactures, is always willing and even solicitous to export its own. The popular notion is, that this is the only way to get any profit out of the foreigner, and that whatever is taken in the shape of merchandise from him, is so far a loss to the home trade. It cannot be too often repeated that we can scarcely ever purchase foreign manufactures without exporting home manufactures as the only means of paying for them. They must be paid for in some way—that is, in money or in goods.

Now in an earlier part of this paper I have tried to prove that large commercial transactions cannot be settled by money, and at the risk of repetition I must dwell upon it a little further; and for this especial reason, that it is from a misconception of the part that money plays in the economy of trade, that so much error on this subject prevails.

It is a common saying of those who only possess a superficial knowledge of the matter, that we are constantly draining the country of specie, in order to pay for the large excess of our imports over our exports, and it is of course assumed that whenever specie leaves this country, we are losers by the event: just as if the owner of the gold or silver would part with it, unless he obtained in return property of a greater value. And besides this, the money itself would probably be the result of home labour, in the shape either of natural produce or manufactures. It is said to be advantageous to export goods, disastrous to export specie. Why? Because, the answer is, the latter would have supported labour. But labour is not nourished and sustained by money, but by what money can purchase. Goods may be exchanged indirectly for sustenance as well as money can.

A certain amount of specie is requisite in every country for the due circulation of the property it possesses. Allowing for ordinary fluctuations, it is on the whole neither more nor less. If there is more than is requisite, its value falls, and it becomes an encumbrance, just as more furniture in a house than can be required is not only in the way, but is an actual nuisance rather than a benefit. If there is too little, the inconvenience is felt in the other direction. Any serious disturbance in the balance between demand and supply would enhance or depress the value of money accordingly, and any large and continuous exportations of it in payment for goods, would be neither expedient nor profitable. I have shown elsewhere that there is a limit to the desires of mankind for mere necessaries, as well as for simple conveniences. No one would wish for more food than he could consume, nor more chairs in his house than could possibly be sat upon; he might have a hankering after more costly provisions or highly decorated furniture, and just so far as they went beyond simplicity or usefulness they would become luxuries, and to these it is clear that there can be no limit. Now money is a convenience, and nothing more, and consequently has its limits in use: just as a particular tool has to a workman. One is as serviceable to him while it lasts, as a dozen of the same kind would be. In the same way money is a mere tool to trade and commerce. It is a means to an end, but not the end itself. No one desires money for its own sake, but merely that it may be the medium for obtaining what may afford him substantial pleasure and gratification, and of these we never feel that we have a superfluity. A collector of pictures or coins or autographs, for example, is constantly craving for more, and only wishes for money as a means of augmenting his store of things which please his fancy.

Money is, in fact, very like the counters used by

gamblers in a game of loo or speculation. They have a certain nominal value, and bear the same relation to the stakes played for, as money bears to merchandise in the ordinary traffic of life. The players are anxious, no doubt, to win them, but merely that they may attain the substantial objects of the game. Large sums of money may pass from one player to the other, but this does not in the least affect the number or value of the counters. If there are enough to carry on the play conveniently, a larger quantity would encumber rather than assist the progress of the game. Money then, constitutes the counters with which men play the game of commercial exchanges of merchandise. The bartering of one species of goods for another is the main object of the play; money, simply as such, is very little concerned with the substantial result.

But to return to our islanders; if A be willing to supply B with corn (which it certainly would be if requested), although it will not take the particular article, cotton, in return, why should not B avail itself of the opportunity, instead of wasting twice the amount of money and labour in raising it for itself? It is true that if B had no other direct mode of paying for the corn except by cotton, which A would not receive, and it could not get it elsewhere, there might be a difficulty, and B must then submit to be its own grower at whatever cost. But such a state of things would be very unlikely to exist. In the first place, there would be many other commodities that were ordinarily exchanged between these two islands, upon which no such tax was imposed; and in the next, islands C, D, E, and F, who were in general course of trade with A and B, as well as with one another, would very likely import corn from A, pay for it in their special productions on which A had created no restriction, and then B might get supplies of corn from them in exchange for its cotton, on the importation of which the other islands might be too well advised to put prohibitory duties. It is in this circuitous way that commercial transactions are continually carried on.

At all events, if the inhabitants of B were to put duties on A's corn, that they might raise it for themselves, when they had the means at hand of obtaining it at half the price it would cost them, they would not be giving much proof of their intelligence. To buy in the cheapest as well as to sell in the dearest market, quite irrespective of the locality whence the commodity or the payment may come, is surely the truest policy with reference to your own population, as it is to every other.

To act as fair traders suggest, would be very much as if, because you had already suffered one infliction at the hands of a foreigner, you would, without obviating the first, voluntarily subject yourself to another, on the simple ground that, although it injured you, it equally injured him.

Now, by way of varying the argument, I will put an illustration in another shape. Suppose the inhabitants of A could buy shoes from B at 5*s.* a pair, whereas similar articles could not be made and sold at A for less than 10*s.* (I select shoes because they stand midway, as it were, between luxuries and absolute necessaries. They are not things which could easily be dispensed with, and at the same time they are not essential to our existence, though every one is desirous of possessing them.) Now suppose that the people of A should take it into their heads that it would be a credit or an advantage to them to have a shoe manufacture of their own, and that they, accordingly, prohibited the importation of shoes from B into their island. The result of that would be that each inhabitant would have to pay 10*s.* for what he could otherwise get for 5*s.* Five shillings would therefore be absolutely sacrificed or lost to him, without any corresponding advantage to set against it, as far as he or any one else was concerned. It would be equivalent to a tax, but of which even the State got no particle.

The shoemaker would not be especially benefited, because he would only get the ordinary profit that was the current one among the artificers in a similar species of trade, and which he could make by his labour in other work for which he was better fitted. If it were found that he got more than the general rate of remuneration for his labour and outlay, competitors would soon spring up, and by underselling him put a stop to his inordinate gain. He would sell his shoes for 10*s*., because he could not afford to sell them for less; and the reason of this would probably arise from the longer time, and consequently the greater amount of labour he had to expend in bringing his work to completion, and this would involve a loss to him of a considerable quantity of food and other necessaries, which must of course be included in the price. Now if perfect free trade were established in A and B, it is evident that this loss might be avoided, and every inhabitant of the former would find himself with an additional 5*s*. in his purse, to be spent in other things, equally the result of labour.

Where then is the advantage to any one of artificially vamping up a particular trade, on the weak

On Wealth and its Sources.

and sentimental pretence that it is right to encourage the home artisan? You may force strawberries to ripen in winter time, and affect to enjoy them at Christmas banquets; but it would be by a very large expenditure of money paid for toil, and after all, the flavour might be far inferior to what nature would provide, if you would be content with her almost spontaneous efforts at production. You might be well able to afford it, and quite justified in doing it, but it would be folly to contend that it was in the end an economical mode of producing the fruit in question.

I think I may truly say, therefore, that what is ordinarily meant by protection to native industry, or the synonyme Fair Trade, is the encouragement to men to enter upon an occupation for which they are comparatively unsuited, and while the rest of the community have to pay the cost of the experiment, the workers themselves are no more benefited than they would be, by any other employment to which they might turn their hands.

There may be a lurking idea in the minds of some people that if this sort of manufacture were not resorted to, those sought to be employed upon it would not get any work at all, and that it is better to set them on partially remunerative toil, than to let them languish in idleness and inaction. Now this is, in the first place, regarding the matter in the light of pure charity. But even then, to recur to the case I have put of the protected shoemaker in island A, probably everybody would be a gainer if part of the money saved by buying shoes in island B, were expended in pensioning off their pet tradesman, and allowing him to live in idleness instead of employing him in wasting half his time by making their shoes. But there is no necessity to resort to the principle of charity at all. I suppose it may be admitted that *prima facie* it would be more beneficial to manufacture those articles in which we excel than those in which we are excelled by others; at least, we should have a larger quantity of commodities than we should otherwise possess. It may be said that possibly there might be no demand for those articles that we had such superior facilities for producing. But you would be destroying all opportunity of testing this, while you were granting monopolies and bounties for the nominal benefit of certain trades, and so almost compelling capital to devote itself to their sickly and unnatural culture. Great Britain has long been considered one of the most favoured nations of the world for the development of invention, industry, and skill in trade and manufactures; in natural productions our soil is rich and prolific in certain cases, as, for instance, in coal and iron. Now if foreign markets were glutted with one species of commodities, they might be open to us in others. English people ought to be the last to admit that they are so far outdone in competition with other countries in manufactures, that they require a sort of body-guard, in the shape of imposts, to protect their interests. At all events it cannot be denied that by fostering illegitimate production, we may be possibly damaging that legitimate enterprise which might in the end be much more profitable, because much more in accordance with the special advantages which in many instances Providence has bestowed upon us.

Is it worth while to run this risk, by authoritatively interfering with a system which has existed for so many years? It seems a singular argument to put forth, that because some countries have done what we all admit is palpably detrimental to their interests, we should, without counteracting the effect of their movement, intentionally subject ourselves to the same evil consequences. There appears, too, to be this further inconsistency in the views of the protectionists. One would suppose that, if they were inclined to carry out the selfish theory of benefiting their own countrymen, regardless of the interests of foreigners, their course would be, churlishly to refuse to the latter, all participation in the special advantages that nature and skill had given them, whether with reference to soil or manufacture; but singularly enough, while they are studiously desirous of extending these benefits universally, they would doggedly refuse to avail themselves of the superiority in this respect that other nations have over them.

That we should give encouragement to our own workmen in preference to those of other countries is, I repeat, a very plausible and very proper contention; but it would be to the last degree impolitic to waste our resources in effecting it. Those who call themselves Fair Traders are in the habit of pointing with great unction to the large excess of our imports over our exports. They allege this to be a great evil, but they do not attempt to enlighten us as to how the imposition of duties upon foreign produce is to remedy the evil, if it really were one.

The figures set before us, however authentic in form, are far from warranting the conclusion that the difference shows the precise extent to which we avail ourselves of foreign labour, over and above what foreigners employ of ours.

I am not going to enter into any detailed statistics as to this matter, because I wish strictly to limit myself to the discussion of abstract principles. I admit that there is a considerable excess of our imports over our exports, and there is an all-sufficient reason why this should be so, without any derogation from the theories I have advanced; and the explanation will be seen presently. But I may say now with reference to this subject, that the published accounts are very misleading, if we seek by their means to elucidate the point we are at present discussing. In the first place, our exports are taken in the tables at their prime cost, without including freight and other expenses, which would, if added, of course considerably swell their amount; while our imports are, on the other hand, reckoned with the accumulated expenditure caused by their transit to this country.

But quite independently of this, the profits of our foreign trade are largely augmented by having in our hands a large preponderance of the carrying trade throughout the world; and this accrues both from exports and imports, though of course not included in the tabulated accounts. They are necessarily excluded from the exports, which are assumed to be in our favour; they are included in the estimate of our imports, and though to our actual profit, help to swell the balance of trade that is alleged to be against us. But the excess of imports, which affords so large a scope for lamentation, and which I have said may be naturally accounted for, without derogation from the argument in favour of free trade, arises in no slight measure from the fact, that a large amount of British capital is invested in enterprises abroad, such as foreign loans, foreign railways, and other like speculations, the interest and dividends on which, must always amount to a very large sum. I hope I have sufficiently established that these cannot be paid in gold and silver coin. They must in general be sent to this country in merchandise, which is sold here, and the proceeds, in the shape of money, paid to the owners of the capital invested. Of course there can be no exports to set off against such imported commodities, because they have been already paid for by the use of the loan in the country whence the goods came.

Now this raises the very important question I have before touched upon, and affords at the outset, an apparently fair topic of complaint to those who honestly sympathize with home artificers and labourers, because it seems to trench upon a principle we all recognize, that we should, in the employment of our capital, give the preference to our own countrymen over strangers. If, in resorting to these foreign investments, we really did prejudice the British workman, I would denounce it as much as any fair trader could do, although, after all our denunciations, it would be difficult, certainly impolitic, to interfere with the liberty every man has, in this direction at least, of dealing with his property in any way he pleases. Assuredly the levying duties on foreign goods would not effect this object. It would, in truth, increase the quantity imported, and so add to the inquietude of the fair trader, since, as the duty must fall ultimately on the person entitled to receive the interest on the investment, a larger quantity of merchandise must be sent to enable him to receive the amount that was due to him.

But further to elucidate this topic, let me enter a little more minutely into the question of capital, which has only been lightly referred to in an earlier portion of my subject. The word is not always used with precisely the same meaning. It must always however be taken to apply to the accumulation of the savings of labour after the remainder of the general produce, of which it once substantially formed part, had been consumed or destroyed in ministering to the necessities or pleasures of individuals, and so of the community at large. And these savings are stored up in the shape of quasi-permanent and tangible property. Any variation in the sense in which the word is used, merely arises from the mode of employment of the capital itself. In the first place there is fixed capital, which means plant, machinery, tools, &c., which are constructed for one single and restricted purpose, and if that purpose be altogether superseded, the subject-matter being available for no other, becomes absolutely valueless. It was of great value while its operations were called for, because it saved the expenditure of labour, or what is the same thing, it manufactured its peculiar objects much more expeditiously than the unassisted toil of man could possibly accomplish, and thus limited the consumption of necessaries which would be expended by the greater number of persons who would otherwise be employed to do the same quantity of work. The stocking-loom, for instance, would produce in a given space of time an infinitely larger quantity of stockings than hand-knitters of themselves could turn out; but if men ceased to wear stockings, the existing looms could be of no use to any one.

Most property is of some value, because if one

man does not require it, another may ; but this is because it is an object of desire and of utility in some way. The stocking-loom, if people gave up wearing stockings, could be an object of desire or utility to no human being.

Another species of capital is circulating capital, which is used in the ordinary transactions of trade and commerce, the substance and foundation of buying and selling, the fund for paying wages and such like purposes. It is constantly changing hands, and is in fact so dealt with, that some profit may arise from exchanging it for something that may be of more value to the dealers, whether in the shape of goods, money, or labour. But from what I have said before, it will readily be understood that though the popular idea of wealth is represented by money, a comparatively insignificant portion of it consists of specie.

Circulating, again, differs from fixed capital in this, that it is nearly always convertible, and therefore valuable, and this on account of the universality of its application.

But there is a third species of capital which is but seldom adverted to, but which forms a vast item in the general available wealth of a nation, and this I will call, for want of a better term, domestic capital. I cannot perhaps give a better idea of what I mean, than by directing attention to one of the fashionable streets at the west end of London, and begging a moment's reflection to the amount of riches enshrined in every dwelling-house there—the costly furniture, the works of art, the plate, the jewellery, and various other articles of luxury that are considered as the ordinary appendages of an opulent household. And if we extended our contemplation on this score, we should find in every dwelling-house throughout the kingdom—of course varying with the station of the occupant—that there is a large amount of superfluous objects of cost and value, which, in case of necessity or strong inducement, might be converted into active means of employing labour and increasing wealth, and this without interfering in the least degree with the substantial convenience or comfort of the owner. But enormous as this stored-up capital must be, it is seldom or never intended originally to be used for any such purpose. It is considered to be almost a necessary appertaining to a man's position, and whatever that position may be, he must needs keep up appearances, as they are called, in the eyes of the world—meaning, in general, those of his associates in the same rank of life as himself. He would no more part with them than—to use a very humble comparison—he would part with his umbrella, his spectacles (if he requires them), or his razor (if he does not prefer a beard to being close shaved), though these would be much more useful.

Now it will be readily seen that this kind of capital differs essentially from both of the others: from fixed capital, that it is always valuable and convertible; and from circulating capital, inasmuch as it remains stationary, and is never involved in trade or speculation, except when temptation in the shape of large profits assails it. It is needless to say that, while it remains in the state I have before described, it is quite unproductive. It neither augments the products of the soil, nor does it in any way employ labour.

Much as I have said about the subordinate part that money plays in the estimate of our national wealth, I must repeat that nothing is more difficult than to dissociate the word in the ordinary mind from the term capital. We constantly hear it said that "a man does not know what to do with his money." This is not quite what is meant, or, at all events, what is correct. Substitute the word property for money, and the observation might be just. A person is said to leave several thousand pounds behind him, while perhaps 100*l.* in hard cash constituted more than the sum total of his possessions in that particular article. The thousands only of course mean, in value or money's worth; and into money, if necessary, it could always be converted: but if the conversion took place in this country, it would be a mere exchange, and would leave the aggregate wealth unaffected. But it might be, that those into whose hands the property came, might see their way to some profitable speculation at home, so that by selling and embarking the proceeds in some novel enterprise, a large gain might accrue to themselves, as well as to the public, by increasing its general funds. As things stood before, the property might have remained dormant, and the money value might have been invested in the same species of articles as, for years, it consisted of, and they would be dormant and unproductive too. To make property breed, as it were, much of course depends upon the energy, skill, and industry of the possessor; and the more it circulates, the greater chance there is of finding a judicious and successful operator. Of two men, one may be wealthy without enterprise; another may be enterprising without wealth to render his talent available. Separate, they both stagnate, and the country is benefited by neither; but by amalgamating their different advantages, large profits might arise—not only individually, but universally—because an increase of wealth in any particular locality, of course increases the wealth of mankind collectively.

It is true that opportunities for brilliant speculations do not often occur in old communities like ours. Great occasions sometimes do spring up, as in the case of railways, gas, and steamboat introductions in former times; and probably electricity will be productive of others in the present generation. We know with what avidity capital of all kinds, which had hitherto lain dormant, rushed into the first-named undertakings, wildly and ruinously to many of the early adventurers, but eventually subsiding into enormously profitable operations.

Yet the current of our ordinary trade and business flows on slowly and cautiously, and offers little inducement to men to run the risk of employing their time and their savings in venturing upon them. I have pointed out before, that wherever capital is great in proportion to the demand for it, its profits are necessarily small. With us, assuming that the bare interest of money is from four to five per cent., many prefer to place their savings in the public funds, where they get little more than three per cent., to running the hazard of losing their principal altogether.

It may not be out of place here to mention that when we talk of putting money in the Public Funds, or in the Stocks, or in Consols, we do not quite correctly state what the transaction consists of. It is sometimes thought that, in such cases, money is taken to the Bank, and there deposited for the use of the Government, which agrees to give a certain annual sum by way of interest for its use. But in truth, Government has nothing to do with the proceeding, except to provide by taxation from time to time the sums accruing due for such interest or dividends. It never touches the principal, for there is no principal to touch. In former and even in recent times, when the State was engaged in onerous wars, or was pressed by other emergencies that required a large and immediate expenditure, it was in the habit of drawing on posterity, by borrowing money from the public, and undertaking to pay for the use of it, by levying in future, taxes on the people generally, until the loans were paid off. The sums so borrowed have very long since been spent and dissipated ; and inasmuch as these accumulated amounts so raised—called the National

Debt—now reach to between 700 and 800 millions sterling, the chance of their ever being repaid is somewhat remote: but the obligation to pay the interest to the successors of those, whomsoever they may be, who originally advanced the principal, remains; and as long as this country continues to enjoy the reputation it now holds throughout the world, the pledge so given will never cease to be redeemed.

It is true that public faith has recently sustained a severe strain—perhaps it may be said, for the first time in the history of the nation—by the mode of dealing with Irish landlords. Many of them purchased land on the distinct guarantee that they should enjoy the full usufruct of the soil without let or molestation—at least from the Legislature: they were assured that they would in fact possess a parliamentary title to that in which they were invited to invest their money. But now, by a second statute, in direct contravention of the former one, they are forcibly deprived of a large portion of the profit they fairly expected to realize, and their rights are as deliberately confiscated as they were before deliberately established. I will not here enter into any controversy as to whether, under the peculiar circumstances of the case, this course was or was not absolutely necessary; but it certainly seems questionable on the score of political morality. It forms a most dangerous precedent, and precedents are things that legislators are apt to quote when they desire to make a tyrannical use of their power, backed as it sometimes is by the still more tyrannical voice of a capricious public. What is to be feared is, that the pernicious principle lately acted on may eventually prove the first step towards national dishonour. It has been said that the measure only affects our own fellow-countrymen, and that inasmuch as the interests of the few must needs give way to the welfare of the many, therefore individual equities ought not to be regarded.

It is a very delicate and specious distinction, and might be resorted to, to sanction the most oppressive infractions of honesty and fair dealing. But injustice is injustice still, whatever may be its scope, or whosoever may be its victims. The light-hearted performers in this game of spoliation may seek to silence the whisperings of conscience by these transparent sophistries, but they leave behind them a rankling sense of wrong in the minds of the people, and do infinite mischief by destroying their confidence in those who are placed in authority over them. But at all events, any such plea as I have referred to would be utterly nugatory in the case of the National Debt, since very many foreigners hold large amounts of stock in what are termed the Public Funds. Repudiation has been resorted to by other states elsewhere, and we may remember what scorn and obloquy were lavished on its perpetrators.

But to quit this digression, when a man is said to invest in the Stocks, it means that he simply purchases from somebody else, the right that other possessed of demanding from the State an annual sum in the shape of interest due on a sum of money long since lent, and not a trace of which remains. There are no other parties to the transaction. Such rights are the subject of continual purchase and sale in the open market. They can always be bought, and always be sold, at one price or another. For there are men who make it their special business to deal in this species of property, and are at all times ready to sell at a trifle somewhat above the market price, and buy at a trifling reduction below it. Inasmuch therefore as there is a constant fluctuation in the number and identity of the owners for the time being, it is of course essential that the Government should know who are entitled to receive the dividends. But this duty is entirely taken off their hands by the Bank of England, which, in consideration of certain ulterior privileges granted to it, keeps an exact register of every transfer that takes place from one person to another, so that there is thus complete evidence of the holder's title always afforded.

Now remember that, with regard to the man who buys and the man who sells, their joint property remains the same as ever. It is a mere exchange between them. The former may be willing to pass a life of ease and unproductive leisure, satisfied with the pittance of three or four per cent., which he is sure can never fail him while the present constitution lasts. But he who receives the money for the stock may be more active and energetic, and be disposed so to use it as to make it produce to him much more than the trifling amount of interest he would get by retaining the security in his hands. A tradesman or a manufacturer with a capital of 1000*l.* might so use it, that by turning it over (as it is called) several times—that is, by making or selling his wares frequently in the course of a year, and getting a profit on each transaction— he might, at the end of it, not only be able to replace the capital, but find himself in possession of an equivalent sum to that with which he started. He would thus have doubled the amount he originally possessed. In the process of effecting this, he would have employed and sustained a large amount of labour, have given a strong impetus to reproduction, and thus have largely increased the wealth of the community among whom he lived. That 1000*l*. which would only have produced to him three or four per cent. had he held it in the public funds, would now return him 100*l*. per cent. The extra gain could not of course be assigned to mere interest on capital. It would be due to the skill, intelligence, and industry of the operator. If it may be urged that I am putting an extreme case, it is at all events one that very frequently occurs. But I am only desirous of showing in what different ways capital may be employed. If the owner had spent his 1000*l*. on plate or sculpture, the whole would have been absorbed in metal and marble, and further reproduction from it would be gone for ever. In saying this, however, I am anxious to guard myself from the imputation of decrying the culture of art in any of its various shapes. All I mean is—and I have suggested this before—that it has a tendency to consume rather than to create substantial produce, and can only be justifiable where there is a superfluity of wealth, after the actual requirements and conveniences of life have been provided for. When these have been amply guaranteed, the encouragement of art is not only justifiable, but in the highest degree praiseworthy: at that period of a nation's career when luxuries become inevitable, as they assuredly do, it becomes one of the most advantageous and ennobling forms of their development.

Times, however, will always occur when commerce becomes lethargic, and enterprise is to some extent paralyzed. But if there were a vast and sudden impetus given to trade, by some new invention or some startling discovery that gave promise of materially affecting the comfort and welfare of society, the profits from the employment of capital in that direction, might be so increased that the interest upon the loan would be much increased also. The demand for realized wealth would become considerably enlarged, and according to the invariable rule its price would rise in proportion. After all, trade depression among artificers, does not entirely proceed from there being too much labour in the market as against the gross amount of capital in the country (for I have shown how enormous and available that amount is), but to the indisposition on the part of those who own it, to risk their funds in undertakings that are likely to be but slightly remunerative. A depression in the labour market often arises from the greed of manufacturers, by their resorting to over-production in times of prosperity and hope. They produce in one year what ought to be spread over two or three, and even if their goods are eventually disposed of at a profit—and they not infrequently remain unsaleable on their hands—the demand for labour receives, for a time, a serious check; work becomes scarce, and wages are reduced. On the other hand, artisans, irritated at what they conceive to be the inordinate profits of their employers, demand larger remuneration than the latter can afford, or at all events, more than they need pay, according to the wages tariff at the time. This leads to strikes and locks-out, which I have before shown to be destructive of the interests of both the masters and the men, and at the same time a direct obstacle in the way of any increase to the world's wealth. Depression in trade too, of course, often arises from bad seasons. We have seen that the augmentation of capital depends upon the annual productions of the earth, and constitutes what remains, after the actual necessities of man have been provided for. Be they small or great, the whole population of a country participates in them more or less. Where the season is disastrous, the share of every individual is necessarily reduced; and this is productive of a twofold injury to the trader and the workman, because he not only loses some portion of what he would get in the general distribution, but his customers, being losers also, have less funds to expend on the articles he deals in or works upon. Consequently there is less demand for labour. The mode in which the upper and middle classes meet a casualty of this kind, is by cutting down their expenditure, and dispensing with many of the little luxuries they had previously indulged in. We do not in general hear of their selling off a picture, or a piece of plate, or furniture, to make up the deficiency of income. It is at this period of failing trade, that theorists and agitators put forth their nostrums and specifics, and try to urge the Legislature to officially prescribe them as infallible remedies—just as if legislative enactments could supply those productions of the soil that nature had for a time peremptorily denied to us. We must bear the burden as we best may, and live in hopes that the bounties of Providence may again be extended to us.

Nothing can be gained in such a case, by seeking to avert our gaze from the true source of the calamity, and the worst mode of dealing with it, is to resort to a meddling, empirical interference with a condition of things that is irreparable.

Having then endeavoured to show that there is always a vast amount of unproductive capital, only waiting for an opportunity of being turned to profitable account, I will now recur to the question of foreign investments. I will premise by saying that Great Britain may be likened to a man who, after a life of toil and labour and industry, had become well stricken in years, and had lost much of the physical power which once distinguished him; but he might still have preserved all his intellectual faculties, all his moral and mental energies in their ancient vigour. If he had a large and young family dependent upon him for food, and could only procure it for them by his own bodily exertions, the share that would fall to each child, would probably prove to be a somewhat scanty one. But if, during his past career, he had been saving and thrifty, and had amassed a fair share of wealth, he would, in spite of his faded strength, have no difficulty in maintaining in comfort and prosperity, all those that had domestic claims upon him.

Now it can scarcely be denied that the vitality of our soil, so far as concerns its powers of reproducing the fruits of the earth—what I will call its life-sustaining power—has been very considerably impaired by time. It is true that England has vast stores of mineral wealth—for instance, coal and iron—hidden away in her bosom, and these have been perhaps more conducive than any others that could be named to her mechanical and trading supremacy : but, although these require labour to develop them and render them available, it is obvious they have nothing to do with natural reproduction. The density of our population, in proportion to the extent of our home territory, is much greater than that of any other European nation, with the exception of Belgium. A large portion of our land has been appropriated to manufacturing and other distinct purposes, and that which could conveniently be devoted to agricultural pursuits, has had its powers and its energies overstrained, in seeking to provide for the ever-increasing number of mouths it has to feed. In this particular therefore, we necessarily show a deficiency in vital tone. But that is very far from affording any evidence of general creative decay. It is quite possible—to take a very extreme case—that a nation which never grew a blade of grass or an ear of corn, might be in a richer and more prosperous condition than any of the countries by which it was surrounded. It might exhibit on the part of its inhabitants generally, so much skill and ingenuity in mechanics, in manufactures, in art, and in taste—in manipulating, in short the raw material sent to it by others, that these others might be willing to pour into it large quantities of their own special produce, receiving back in exchange, a small portion in its improved condition, and leaving the remainder as a remuneration for the augmented value of that which was returned to them.

The rough block of marble which contained within it the statue of the Venus de Medici would be of little importance in itself, when sent from Paros to Athens; but after the figure's surroundings had been chiselled away from it by the genius and talent of the artist, it became of priceless worth. What would the canvas and paint of a picture by Raphael be estimated at, before the master had converted them into a thing of beauty that has enthralled the senses of successive generations of men? Genius, invention, and ability, employed in these and other ways by any one nation, might command a much larger quantity of necessaries from the rest, than would be sufficient to supply with food its entire population. Our country has been sometimes called the workshop of the world; others might well be termed the world's granaries; and where there was universal freedom of trade, both must profit by the interchange of those commodities that each had a special aptitude for producing.

Great Britain being thus situated, with her reproductive powers lessened, but with great hoarded-up wealth, there are other countries in a reversed position. They have great internal resources, but they have no funds to develop them, and unless some other nation stepped in to their assistance, to the advantage of course of both, their hidden treasures and unused capabilities would be lost to the world. They are naturally willing to pay a large price for the accommodation rendered to them, and this is the bait that attracts some of the superfluous capital of Englishmen to their territories, and unless so employed it might perhaps never be applied to the maintenance of labour here, as we shall presently see. Take the case, for instance, of a man who possesses a gallery of pictures worth a hundred thousand pounds. The pictures are pleasant to look at, and form a great ornament to his establishment. They constitute realized capital, and are valuable as exchangeable property; but they contribute nothing to production nor to the support of toil. Now assume that for the purpose of assisting the development of wealth in a distant country, where capital was needed, he were to part with these paintings, and invest the proceeds in the furtherance of this object, with the assurance that the speculation would return him ten per cent. dividend. It would be useless sending the pictures to the scene of projected operations; it would be like sending jewellery and trinkets to a population that was dying for want of food. He might, for instance, sell his collection in France or elsewhere. As we have seen, the payment would not be likely to be made in money, but in goods, sent to this country, sold here, and with the sum realized he might purchase machinery, tools, and implements for the purpose of carrying out the foreign improvements. Now to stop, for a moment, to answer some captious objector who might urge that the process I have described would involve the very injustice I have all along admitted, of the employment of foreign labour to the exclusion of that of our own countrymen, as to the merchandise imported here for the payment of the pictures. If the gallery of paintings remained here, simply to bedeck the establishment of the owner, no such merchandise (clearly forming an accession to our aggregate wealth), would ever have reached us at all. We should not have manufactured it ourselves, because, the capital which would have furnished the means of so doing, would have been locked up in a totally valueless and unproductive condition. I suppose the staunchest Fair

Trader would scarcely contend, that, if the Frenchmen were to send us over a large quantity of goods for nothing, it would be right to reject the boon, because the goods were of foreign make, and so deprived our own artificers of the opportunity of earning money by manufacturing them. Yet virtually, in the case I have put, the transaction would be equivalent to their presenting them to us gratuitously. The owner of the pictures, without the temptation of large and unwonted interest, would probably never have drawn his property from its listless, profitless seclusion. The gallery would have remained in its former position, and labour would not have been called into requisition.

Now if it would be wrong to refuse a gift of merchandise from a foreign country, were it kind enough to present us with it, surely I may ask, why should we not buy goods at a cheaper rate rather than make them ourselves at a dearer one, and that irrespective of the market in which the purchase was made? The two cases are the same in principle, though no doubt they differ in degree. For instance, if for a certain sum we can procure twice the quantity of goods from abroad that we can make them for by the same expenditure at home, it is much the same thing as if we obtained the half of them for nothing. We should save the necessaries that would be expended in making the larger quantity, thus reserving it for use in various profitable ways.

At all events, the sale of these hitherto unproductive pictures would, on my assumption, bring into this country 10,000*l*. a year in the shape of interest where nothing came into it before. Now, I have laboured to little purpose, if I have not clearly established that this income could never be enjoyed by its possessor, dispose of it how he might, without being made subservient to the nourishment, the maintenance, and comfort of a large number of persons besides himself. It might be made productive in an agricultural sense, and by profuse expenditure in draining and manuring at the outset, acres of ground might be made to produce crops that it had never yielded before. It might be used in breeding herds of cattle; it might be employed in trade, thus giving a stimulus to production in an indirect way. Even if it were spent by its owner in the mere gratification of his whims and pleasures, many people must be employed in ministering to them, and thus gaining their livelihood.

But it may be fair to assume that, inasmuch as

gain was his object in parting with his pictures, he would endeavour to make the most of the interest resulting from the speculation. Remember that with this 10,000*l.* a year he might employ 200 workmen continuously, at 50*l.* a year each, by their toil and labour, to manufacture for him what would be of sufficient value to replace the whole sum expended upon them. Two hundred families might thus be supplied with at least mere necessaries, and what they had consumed and destroyed would be restored as property in some other shape, thus affording the means of repeating the same process from year to year. Then again, these very families must resort to butchers and bakers, to tailors and shoemakers, and other artificers, to satisfy their requirements, who would all seek some share of the gross profit that this 10,000*l.*, judiciously laid out, had produced. This would be the result of profitably employing hitherto unused wealth, and this, in spite of its being devoted in the first instance to that much-decried object of employing foreign labour.

But even the advantages I have described might not be the full extent of the benefit accruing to us; for the country to whom the capital was originally lent might become so prosperous, by means of the use to which such loan, together with similar ones, had been put, that it might in course of time furnish additional markets for our manufactures, and so open up new sources of encouragement to our native industry.

Having thus shown how stored-up wealth may be called into active service when a sudden demand for it arrives, and an unusually large profit from it is to be made, I need simply refer to the converse proposition when competition slackens, markets become dull, and capital therefore is little in request, consequent on the slender remuneration its employment would afford. Its possessors would then lay it up in ordinary, as it were, by investing it in works of art—pictures and sculptures, for instance (to resort to the same examples as I have before adopted). When so spent, and these articles were newly created, it would have made its last effort at reproduction; for although what was thus made would always command a certain price, it must be recollected that this would only represent the large amount of necessaries absolutely lost to the country in producing them.

Pictures or sculpture will neither multiply nor grow; they will not produce, nor assist in producing, anything beyond themselves, as seeds, machinery, and tools would necessarily do. And although their value might be at any time realized by a sale here, there would even then be a mere substitution of one owner for another without any increase in the things exchanged. But it might happen that, for the moment, the community was so well stocked with commodities that were susceptible of reproduction, that there appeared to be no advantage in increasing them; then would come the period (or even, as I have explained before, the necessity) for resorting to investment in luxuries, and those just referred to, would at all events preserve to their owner something of an exchangeable value, that would be permanent and enduring. Contrast this mode of dealing with superfluous wealth with that of employing it in the manufacture of expensive wines, which would be destroyed in their momentary enjoyment, and the distinction will be at once apparent.

Now this is only carrying out the principles I have all along sought to enforce, that necessaries are the first species of produce of which in an early stage of society a supply must be secured. Then come the conveniences and comforts of life, and when all these have been sufficiently and amply provided for, the introduction of luxuries becomes inevitable, if men are to continue to exercise the faculties Providence has bestowed upon them. And further, that luxuries differ materially among themselves, in respect of the way in which capital is absorbed by them.

Some are created simply to be destroyed in carrying out the intention which gave them birth, without leaving a vestige of result behind. Others, though henceforth utterly unproductive, still endure and last, having always, for the most part, an exchangeable value which may be turned to account when a favourable opportunity occurs.

I am conscious that there is much repetition in what I have lately said, and if this is urged against me, I fear I only strengthen the accusation when I repeat what I said at the beginning, that I would rather incur the charge of redundancy, than that of not having rendered myself thoroughly intelligible. The relations between capital and labour, prices and profits, the necessaries of life and those objects that simply tend to its enjoyment, are not readily appreciated on a first contemplation. But I believe that any difficulty that may exist, arises rather from the fact that men's minds are in general not directed to these subjects, than from any intrinsic obscurity in the subjects themselves. Few persons unacquainted with mathematics would on a first perusal thoroughly comprehend the demonstration of the forty-seventh problem of Euclid. Many would be apt to forget the first part of the proof when they got further on, and would perhaps have to recur to it more than once before they could cordially assent to the truth of the conclusion. I am very far from assuming for the views I have put forth, any claim to a demonstration of their correctness; I only wish to explain my practice of availing myself of perhaps an oft repeated phrase where I think it appropriate in illustrating a new position.

At all events I shall think myself amply repaid for any labour I have taken, if I have counteracted the impression often entertained, or at least sought to be instilled into the minds of the working classes, that the rich man gets all the enjoyment that is to be derived from wealth, and that those whom he employs, do nothing but toil, in order to contribute to his ease and pleasure. If they toil, they do so for their own peculiar emolument and enjoyment as well as for his. But for him, they would in vain seek for the comfort or even sustenance by which they live, in that particular station that is marked out for them, and they can hardly expect that a special dispensation should be provided in their favour, for the purpose of bettering their condition.

They are dazzled by the glare and glamour that surround wealth and high station, and which render them blind to the fact, that a thing which they believe would be a blessing to *them*, may be nothing of the kind to him who is in possession of it. We know that habit and necessity will reconcile us to what formerly we regarded as a most unpleasant and irksome duty. Just in the same way use and custom considerably reduce the value of attractions that once appeared lasting and supreme.

The poor only see the bright side of the picture; they are not aware of the cares, the anxieties, and the responsibilities that are depicted on the other.

After all, happiness is the professed object of every man's desires, and it is quite immaterial by what means (always assuming they are honourable ones) it is achieved. A beneficent Providence has certainly ordained that it shall not depend upon the adventitious attributes of rank or riches, but upon the peculiar condition of each individual mind.

Health and contentment, which may fall quite as much to the lot of the poor as of the rich, are the main ingredients that constitute man's felicity; and if he is fairly endowed with these, there is not on the face of the earth one of his fellow-creatures, whose position he need envy. But a desire to improve his worldly status, so far from acting as a check on the calm serenity of his career through life, is the very stimulus that induces him profitably and pleasurably to employ his energies and his time in attaining that end. Certainly he will never further it by looking with jealousy and ill-will upon those who appear to be more favoured by fortune than himself, and in treating as his enemies those who, perhaps unconsciously to themselves, are his most sterling and substantial friends.

Before I conclude this somewhat desultory and discursive essay, I wish to say a few words on the present aspect of the Free Trade Question.

For upwards of thirty years from the passing of the Corn Law Bill the subject was allowed to rest in peace, and not a symptom of clamour or agitation appears to have been manifested against the new system that had been inaugurated.

This was to be fairly expected when it was once finally established, considering the nature of the opposition that had in a great measure vigorously resisted the change. Very many of those who denounced it were thoroughly imbued with the conviction that Free Trade would ultimately prove highly beneficial to the country. But they honestly believed that suddenly to reverse a policy that had so long been maintained, would be not only infinitely injurious to what were (almost in derision) termed vested interests, but would involve a direct breach of equity and fair dealing. Men had directed their capital into various channels, on the faith that they would not be interfered with by any abrupt onslaught upon their property. Agriculturists would of course be the chief victims, and considering the enormous amount of money that was invested in the land, and the vast number of labourers dependent upon it, the prospective ruin of so extensive a body of the population was not to be lightly disregarded.

But there was a still stronger reason, perhaps, that operated on the minds of many—who were far from being unfavourable to the principle of free importation—in opposing the repeal of the Corn Laws.

The most violent and clamorous advocates of this repeal were the mill and factory owners, with whom trade had not been flourishing of late. They were suddenly seized with a spasmodic fit of patriotism, and a deep feeling of sympathy for what they called the starving labouring classes. Their sensibilities were awakened by seeing so many operatives out of employ, and they displayed their humanity and public spirit by urging and inciting the mobs in the various manufacturing towns blindly to clamour for cheap bread, and to demand with threats, the abolition of the law that prevented the free introduction of foreign corn into this country. The disturbances—the riots, almost amounting to insurrection, that resulted from this beneficent teaching, have become matters of history.

The Abolition Bill passed, and the manufacturing hierarchy reaped their reward, which was by no means a light one, though not quite in the direction they had openly professed to anticipate. They had, of course, predicted that a vast quantity of foreign corn would pour into this country, but they did not so outwardly proclaim, that if this were so, a large quantity of their manufactured goods must necessarily go out of it, by way of payment; for they had diligently studied political economy, and knew that the corn could not be paid for in money. Much of the distress in the factory districts had been caused by over-production, and a large amount of cotton, woollen, and other goods remained on their hands without either home or distant markets to absorb them, even at a sacrifice. These would now be easily disposed of, and their mills and machinery which were a dead loss to them while unused, would be again called into requisition to produce more. These benevolent capitalists were wise in their generation. But the large increase in their trade was not the only source of pecuniary profit they looked to from the change. They had held up landlords and agriculturists to general execration, and compared them to the tax-gatherers and publicans of old, who restricted the food of the people, and battened on the miseries of the poorer classes; but they quite ignored the existence of any other labourers or workmen than their own. It was, however, pretty obvious that if the population was to be fed by foreign corn, large tracts of land which could not compete with foreign produce, must be thrown out of cultivation, and of course it followed that the worst species of soil would be the first to be abandoned, and these were just the farms that demanded the greatest quantity of labour. All those men who had been engaged on them would be thrown out of employ. The land which still remained in tillage, being already well stocked with hands, would require no more. But the discharged men not liking the prospect of starvation, and finding it difficult to accommodate themselves to other occupations, for which their habits had unfitted them, would be willing to work on the only terms on which they could obtain employment, namely, by underbidding their more fortunate brethren who still remained cultivators of the soil.

A general competition takes place, and the wages of agricultural labour inevitably fall. A large multitude still remain to be provided for, and who must exert their industry elsewhere. Their only refuge is the manufacturing towns, there to compete for work with the pet operatives, for whose welfare the mill-owners had all along shown so much amiable solicitude.

Unused to the noxious atmosphere, and bewildered amidst the buzzing of machinery, some lucky individuals among the bucolic emigrants would be soon carried off by disease, or by, to them, the novel eccentricities of cranks, cog-wheels, and other such gear. But the survivors would soon begin to enter into competition with the trained artisans. The same process takes place as previously reduced wages in the country, and the same result assuredly follows. Wages of labour therefore almost universally fall.

Now what effect would this result have on the position of the master? Let us see. I have before referred to the undeniable axiom that the price of goods depends in general upon the quantity of labour employed in their production, and not on the casual rise or fall of the wage that is paid for it. Even McCulloch, the great champion of Corn Law abolition, lays it down broadly, that a diminution in wages will not compel the manufacturer to sell his articles at a cheaper rate, but that the difference finds its way absolutely into his pocket. In exact proportion therefore as the labourer suffers, are the profits of the master enhanced. So that while the factory capitalist gets all the advantage of the increased trade, he obtains in addition, the benefit from each individual sale, by engaging labour at a lower cost.

Now I must not be supposed to lay down these results as permanent ones; that would be to assail every principle that I have sought to maintain throughout these pages. Things would right themselves in time, and an additional impetus would be eventually afforded to trade and commerce. But the immediate consequences would be such as I have described. It frequently happens at first that these violent revulsions have the effect of making the rich richer, and the poor poorer; and the simple reason is, that capital accommodates itself to such changes much more readily than labour possibly can do.

Again, I am far from suggesting that when a great benefit is conferred upon society, the men who have promoted it are to be deprived of all credit, merely because the result accords with their own private interests. But when we see them constantly posing before the world as great public benefactors, and seeking to extort a show of gratitude, where credit for a lucky inspiration is all they are entitled to, and which would be cheerfully conceded to them, it may not be unfair to inquire whether their motives were quite as pure and disinterested as they would seek to represent them. In what I have thus said, I have only been anxious to show that a great part of the resistance given to the abolition of the Corn Laws arose from honest indignation on the part of many who were at heart free traders, to the illicit and unscrupulous means resorted to by these pseudo-philanthropists to bring the scheme to a successful issue. At all events, we of the present generation, possess the advantage of a largely increased system of Free Trade. A long period of time separates us from the injustice and the ruin to many by which the original measure was consummated. A welcome result has been achieved, and I trust and believe we shall continue to retain it. Those who have just started an agitation for a recurrence to the old system, under the plausible and seductive titles of Fair Trade and Protection to Native Industry, mainly consist of a few free lances who are fond of skirmishing on the outskirts of political life. The movement is adopted by neither of the two great parties in the State, nor is it encouraged by any appreciable display of popular feeling.

But if this project is still to be upheld, at least it behoves those who honestly, however mistakenly, advocate it, to state in very specific terms on what ground their hostility to Free Trade is based, because the proposition may be so put, that the most inveterate Free Trader might hesitate to give a categorical opinion on the subject. We are entitled to ask for openness and candour, inasmuch as they are not very conspicuous among public men of the present day. There is a tendency on the part of many to appear to address an assembly in one sense, when they know perfectly well they will be understood by a large portion of their audience in a totally different one; and no doubt great encouragement has been given to the practice, by the example of some in very high places. The art of introducing ambiguity into political speeches is well worth cultivating, by those who are not overburdened with conscientious scruples as to the means by which they attain their ends. It enables them so to frame a statement, that though at first sight it bears one simple and obvious meaning, it may by microscopical search, and by the aid of a little casuistry by way of explanation, be made susceptible of several different and even discordant ones; and it has this great advantage, that the speaker may, on an emergency, adopt that particular construction that is most convenient to him at the moment. The design usually succeeds, when skilfully and carefully executed, because the courtesies of political warfare compel us to admit that a man must himself be the very best exponent of what were his intentions; and we, outwardly at least, give him credit for sincerity, however persuaded we may be of its utter absence. It would be rude to express a doubt where we cannot demonstratively refute. It has, however, a debilitating effect on a man's reputation, if he make these drafts upon public credulity too frequent.

These remarks I conceive are peculiarly applicable to the question of Free Trade, because there is no political topic that affects a larger variety of people in more various ways. By judiciously selecting his audience, and adapting his discourse to its peculiar and well-known wishes, an orator might secure a favourable hearing from twenty different sections of the community, all of which differed essentially among themselves, each supposing that what might be advantageous to itself must necessarily be advantageous to the whole community.

In this selfish age every one wants protection with respect to his own individual trade, but strongly insists on free trade with regard to every other. The motive is quite transparent and not unnatural; they want to secure to themselves a monopoly that they may sell their wares at a high rate, but they are equally anxious to buy everything they require at a low one. Farmers would ask for a duty on foreign corn, that they might obtain an exclusive market for their own produce, but they would resist a prohibition of French silks, lest they should have to pay more for the article than was necessary; but the silk manufacturer, on the other hand, would be in favour of it, while he would clamour against a duty on foreign corn, because he thought the welfare of the country depended on cheap bread. This is precisely in accordance with the popular doctrine of liberty, as its principle is now in general understood; its formula as propounded and acted upon in many quarters being this—"I have a perfect right to interfere with everybody, but let everybody beware how he interferes with me."

To impose a tax upon all foreign commodities would be to partially reduce the nation to the primitive condition of producing for itself everything it required for use and enjoyment, whatever cost it might entail. Whereas to make a selection, and tax only a portion of them for the ostensible benefit of the makers and their workmen, would be to compel the rest of the community to pay an extravagant price for the productions of the favoured few. Even this last expression is scarcely a correct one, for I hope I have satisfactorily proved that the very men for whose supposed protection the duties were imposed, would find that they had gained no boon at all, since competition would soon reduce their profits to the ordinary and normal standard.

It is sometimes confidently asserted, as though the statement must carry conviction to every rational mind, that the staple trades of our country should be maintained at all risks, since they necessarily give employment to so large a body of our countrymen. But it must be remembered that the staple trades of our country cease to be such, at all events in the sense in which the term is used, when we can be undersold by other nations in the subject-matter of such trades. It might just as well be asserted, that we are bound to support in his particular business some large employer of labour in London, for fear of throwing out of occupation, the great number of persons he had engaged in his service; and this, notwithstanding we could obtain articles of his manufacture at a much less price in a neighbouring county than he could afford to make them at. The community generally can have no interest in especially fostering any particular branch of trade or commerce more than any other. The whole nation is composed of individuals, and if it is to the advantage of one man to buy what he may require as cheaply as he can, which it manifestly is, it is equally clear that it is advantageous to all. We need not shed a tear over the annihilation of a class of industry that cannot successfully sustain a competition with other people who embark in it.

I must not omit to recur here (and I wish to do so specifically) to an argument sometimes used in favour of imposing duties upon foreign produce, and which appears to me to be the only one which has a shadow of reason to support it. It is thought by many—as well Free Traders as Fair Traders—that it might be expedient at this juncture to prohibit to some extent the importation of commodities from those nations that levy foreign duties upon ours, on the ground that it might be a means of coercing them into a withdrawal of the restrictions that they have so wantonly and so foolishly prescribed. Now, if this is recommended with that single object, and with the idea that by making a considerable pecuniary sacrifice at the present moment, we might secure great advantages hereafter, the scheme is at least intelligible, and might afford fair matter for discussion. I fear, however, that in the present temper of European and other countries, it would be altogether hopeless. It would involve many hazardous changes in the distribution of capital now, and many indefinite and dangerous ones in future. This would necessarily be so, even though it should eventually turn out to be successful; while the very uncertainty of the result would produce great derangement in commercial affairs for a long time to come.

But I am content to have barely alluded to the subject, and shall say no more respecting it, and for this simple reason, that it is quite beyond the scope of my views as expressed throughout these pages.

The suggestion admits that there would be a sacrifice of substantial wealth by enforcing duties on foreign productions, though it might be desirable to submit to it for some ulterior purpose. This I have no interest nor motive here in opposing, however visionary I may deem the expectation of any benefit to be derived from the experiment. All I contend against is the fallacy put forth by Protectionists and others, that the course they suggest would give valuable encouragement to native industry, and tend to the immediate pecuniary advantage of the whole community.

It is to be feared as much as it is to be deprecated that the movement is taken up as a mere political weapon well calculated to provoke agitation amongst those who are assiduously taught to believe that they would be vast gainers by the change.

The cry is an alluring one, and might perhaps catch some few votes at the hustings; but it is scarcely creditable to the intellect or patriotism of those who resort to it for such a purpose.

It may be considered to be a strong expression of partisanship—perhaps it may be taken to be a sign of dotage on the part of one who has passed through many years of life,—but I nevertheless have always had a strong conviction that a large section of the Liberal party affect to patronize the working classes, whilst they play upon their weaknesses. The Conservatives really seek to protect them, by awakening them—rather roughly perhaps at times—to a sense of their true interests. But, if the movement in favour of Protection had gone further than it has done, I should have begun to think—considering the quarter whence it emanates—that the principles, or at least, the tactics of the Whigs and the Conservatives were, in point of morality, much upon a level.

THE END.

LONDON :
PRINTED BY GILBERT AND RIVINGTON, LIMITED,
ST. JOHN'S SQUARE.

1

A *Catalogue of American and Foreign Books Published or Imported by* MESSRS. SAMPSON LOW & CO. *can be had on application.*

Crown Buildings, 188, *Fleet Street, London,*
November, 1882.

A Selection from the List of Books

PUBLISHED BY

SAMPSON LOW, MARSTON, SEARLE, & RIVINGTON.

ALPHABETICAL LIST.

A CLASSIFIED *Educational Catalogue of Works* published in Great Britain. Demy 8vo, cloth extra. Second Edition, revised and corrected, 5*s.*

About Some Fellows. By an ETON BOY, Author of "A Day of my Life." Cloth limp, square 16mo, 2*s.* 6*d.*

Adams (C. K.) Manual of Historical Literature. Crown 8vo, 12*s.* 6*d.*

Adventures of a Young Naturalist. By LUCIEN BIART, with 117 beautiful Illustrations on Wood. Edited and adapted by PARKER GILLMORE. Post 8vo, cloth extra, gilt edges, New Edition, 7*s.* 6*d.*

Alcott (Louisa M.) Jimmy's Cruise in the "Pinafore." With 9 Illustrations. Second Edition. Small post 8vo, cloth gilt, 3*s.* 6*d.*

—— *Aunt Jo's Scrap-Bag.* Square 16mo, 2*s.* 6*d.* (Rose Library, 1*s.*)

—— *Little Men: Life at Plumfield with Jo's Boys.* Small post 8vo, cloth, gilt edges, 3*s.* 6*d.* (Rose Library, Double vol. 2*s.*)

—— *Little Women.* 1 vol., cloth, gilt edges, 3*s.* 6*d.* (Rose Library, 2 vols., 1*s.* each.)

—— *Old-Fashioned Girl.* Best Edition, small post 8vo, cloth extra, gilt edges, 3*s.* 6*d.* (Rose Library, 2*s.*)

—— *Work, and Beginning Again.* A Story of Experience. (Rose Library, 2 vols., 1*s.* each.)

—— *Shawl Straps.* Small post 8vo, cloth extra, gilt, 3*s.* 6*d.*

—— *Eight Cousins; or, the Aunt Hill.* Small post 8vo, with Illustrations, 3*s.* 6*d.*

—— *The Rose in Bloom.* Small post 8vo, 3*s.* 6*d.*

—— *Under the Lilacs.* Small post 8vo, cloth extra, 5*s.*

A

Alcott (Louisa M.) An Old-Fashioned Thanksgiving Day. Small post 8vo, 3s. 6d.

—— *Proverbs.* Small post 8vo, 3s. 6d.

—— *Jack and Jill.* Small post 8vo, cloth extra, 5s.

"Miss Alcott's stories are thoroughly healthy, full of racy fun and humour . . . exceedingly entertaining . . . . We can recommend the 'Eight Cousins.'"—*Athenæum.*

Aldrich (T. B.) Friar Jerome's Beautiful Book, &c. Very choicely printed on hand-made paper, parchment cover, 3s. 6d.

—— *Poetical Works. Édition de Luxe.* Very handsomely bound and illustrated, 21s.

Alford (Lady Marian) See "Embroidery."

Allen (E. A.) Rock me to Sleep, Mother. 18 full-page Illustrations, elegantly bound, fcap. 4to, 5s.

American Men of Letters. Lives of Thoreau, Irving, Webster. Small post 8vo, cloth, 2s. 6d. each.

Ancient Greek Female Costume. By J. MOYR SMITH. Crown 8vo, 112 full-page Plates and other Illustrations, 7s. 6d.

Andersen (Hans Christian) Fairy Tales. With 10 full-page Illustrations in Colours by E. V. B. Cheap Edition, 5s.

Andres (E.) Fabrication of Volatile and Fat Varnishes, Lacquers, Siccatives, and Sealing Waxes. 8vo, 12s. 6d.

Angling Literature in England; and Descriptions of Fishing by the Ancients. By O. LAMBERT. With a Notice of some Books on other Piscatorial Subjects. Fcap. 8vo, vellum, top gilt, 3s. 6d.

Archer (William) English Dramatists of To-day. Crown 8vo, 8s. 6d.

Arnold (G. M.) Robert Pocock, the Gravesend Historian. Crown 8vo, cloth. [*In the Press.*

Art and Archæology (Dictionary). See "Illustrated."

Art Education. See "Illustrated Text Books," "Illustrated Dictionary," "Biographies of Great Artists."

Art Workmanship in Gold and Silver. Large 8vo, 2s. 6d.

Art Workmanship in Porcelain. Large 8vo, 2s. 6d.

Artists, Great. See "Biographies."

Audsley (G. A.) Ornamental Arts of Japan. 90 Plates, 74 in Colours and Gold, with General and Descriptive Text. 2 vols., folio, £16 16s.

Audsley (W. and G. A.) Outlines of Ornament. Small folio, very numerous Illustrations, 31s. 6d.

Auerbach (B.) Spinoza. 2 vols., 18mo, 4s.

Autumnal Leaves. By F. G. HEATH. Illustrated by 12 Plates, exquisitely coloured after Nature; 4 Page and 14 Vignette Drawings. Cloth, imperial 16mo, gilt edges, 14*s.*

*B*ANCROFT (G.) *History of the Constitution of the United States of America.* 2 vols., 8vo, 24*s.*

Barrett. English Church Composers. Crown 8vo, 3*s.*

THE BAYARD SERIES.
Edited by the late J. HAIN FRISWELL.

Comprising Pleasure Books of Literature produced in the Choicest Style as Companionable Volumes at Home and Abroad.

"We can hardly imagine better books for boys to read or for men to ponder over."—*Times.*

Price 2*s.* 6*d.* each Volume, complete in itself, flexible cloth extra, gilt edges, with silk Headbands and Registers.

The Story of the Chevalier Bayard. By M. De Berville.
De Joinville's St. Louis, King of France.
The Essays of Abraham Cowley, including all his Prose Works.
Abdallah; or, The Four Leaves. By Edouard Laboullaye.
Table-Talk and Opinions of Napoleon Buonaparte.
Vathek: An Oriental Romance. By William Beckford.
The King and the Commons. A Selection of Cavalier and Puritan Songs. Edited by Professor Morley.
Words of Wellington: Maxims and Opinions of the Great Duke.
Dr. Johnson's Rasselas, Prince of Abyssinia. With Notes.
Hazlitt's Round Table. With Biographical Introduction.
The Religio Medici, Hydriotaphia, and the Letter to a Friend. By Sir Thomas Browne, Knt.
Ballad Poetry of the Affections. By Robert Buchanan.
Coleridge's Christabel, and other Imaginative Poems. With Preface by Algernon C. Swinburne.
Lord Chesterfield's Letters, Sentences, and Maxims. With Introduction by the Editor, and Essay on Chesterfield by M. de Ste.-Beuve, of the French Academy.
Essays in Mosaic. By Thos. Ballantyne.
My Uncle Toby; his Story and his Friends. Edited by P. Fitzgerald.
Reflections; or, Moral Sentences and Maxims of the Duke de la Rochefoucauld.
Socrates: Memoirs for English Readers from Xenophon's Memorabilia. By Edw. Levien.
Prince Albert's Golden Precepts.

A Case containing 12 Volumes, price 31*s.* 6*d.*; or the Case separately, price 3*s.* 6*d.*

Beaconsfield (Life of Lord). See "Hitchman."
Begum's Fortune (The): A New Story. By JULES VERNE. Translated by W. H. G. KINGSTON. Numerous Illustrations. Crown 8vo, cloth, gilt edges, 7*s.* 6*d.*; plainer binding, plain edges, 5*s.*

Ben Hur: A Tale of the Christ. By L. WALLACE. Crown 8vo, 6s.

Beumers' German Copybooks. In six gradations at 4d. each.

Bickersteth's Hymnal Companion to Book of Common Prayer may be had in various styles and bindings from 1d. to 21s. Price List and Prospectus will be forwarded on application.

Bickersteth (Rev. E. H., M.A.) The Clergyman in his Home. Small post 8vo, 1s.

———— *The Master's Home-Call; or, Brief Memorials of Alice* Frances Bickersteth. 20th Thousand. 32mo, cloth gilt, 1s.

———— *The Master's Will.* A Funeral Sermon preached on the Death of Mrs. S. Gurney Buxton. Sewn, 6d.; cloth gilt, 1s.

———— *The Shadow of the Rock.* A Selection of Religious Poetry. 18mo, cloth extra, 2s. 6d.

———— *The Shadowed Home and the Light Beyond.* 7th Edition, crown 8vo, cloth extra, 5s.

Biographies of the Great Artists (Illustrated). Crown 8vo, emblematical binding, 3s. 6d. per volume, except where the price is given.

Claude Lorrain.*
Correggio, by M. E. Heaton, 2s. 6d.
Della Robbia and Cellini, 2s. 6d.*
Albrecht Dürer, by R. F. Heath.
Figure Painters of Holland.
Fra Angelico, Masaccio, and Botticelli.
Fra Bartolommeo, Albertinelli, and Andrea del Sarto.
Gainsborough and Constable.
Ghiberti and Donatello, 2s. 6d.
Giotto, by Harry Quilter.
Hans Holbein, by Joseph Cundall.
Hogarth, by Austin Dobson.
Landseer, by F. G. Stevens.
Lawrence and Romney, by Lord Ronald Gower, 2s. 6d.
Leonardo da Vinci.
Little Masters of Germany, by W. B. Scott.
Mantegna and Francia.
Meissonier, by J. W. Mollett, 2s. 6d.
Michelangelo Buonarotti, by Clément.
Murillo, by Ellen E. Minor, 2s. 6d.
Overbeck, by J. B. Atkinson.
Raphael, by N. D'Anvers.
Rembrandt, by J. W. Mollett.
Reynolds, by F. S. Pulling.
Rubens, by C. W. Kett.
Tintoretto, by W. R. Osler.
Titian, by R. F. Heath.
Turner, by Cosmo Monkhouse.
Vandyck and Hals, by P. R. Head.
Velasquez, by E. Stowe.
Vernet and Delaroche, by J. R. Rees.
Watteau, by J. W. Mollett, 2s. 6d.*
Wilkie, by J. W. Mollett.

* *Not yet published.*

Bird (H. E.) Chess Practice. 8vo, 2s. 6d.

Birthday Book. Extracts from the Writings of R. W. Emerson. Square 16mo, cloth extra, numerous Illustrations, very choice binding, 3s. 6d.

———— *Extracts from the Poems of Whittier.* Square 16mo, with numerous Illustrations and handsome binding, 3s. 6d.

*Birthday Book. Extracts from the Writings of Thomas à
Kempis.* Large 16mo, red lines, 3s. 6d.
Black (Wm.) Three Feathers. Small post 8vo, cloth extra, 6s.
—— *Lady Silverdale's Sweetheart, and other Stories.* 1 vol.,
small post 8vo, 6s.
—— *Kilmeny: a Novel.* Small post 8vo, cloth, 6s.
—— *In Silk Attire.* 3rd Edition, small post 8vo, 6s.
—— *A Daughter of Heth.* 11th Edition, small post
8vo, 6s.
—— *Sunrise.* Small post 8vo, 6s.
Blackmore (R. D.) Lorna Doone. Small post 8vo, 6s.
—— *Edition de luxe.* Crown 4to, very numerous Illustrations, cloth, gilt edges, 31s. 6d.; parchment, uncut, top gilt, 35s.
—— *Alice Lorraine.* Small post 8vo, 6s.
—— *Clara Vaughan.* 6s.
—— *Cradock Nowell.* New Edition, 6s.
—— *Cripps the Carrier.* 3rd Edition, small post 8vo, 6s.
—— *Mary Anerley.* New Edition, small post 8vo, 6s.
—— *Erema; or, My Father's Sin.* Small post 8vo, 6s.
—— *Christowell.* Small post 8vo, 6s.
Blossoms from the King's Garden: Sermons for Children. By
the Rev. C. BOSANQUET. 2nd Edition, small post 8vo, cloth extra, 6s.
Bock (Carl). The Head Hunters of Borneo: Up the Mahakkam, and Down the Barita; also Journeyings in Sumatra. 1 vol.,
super-royal 8vo, 32 Coloured Plates, cloth extra, 36s.
Bonwick (James) First Twenty Years of Australia. Crown
8vo, 5s.
—— *Port Philip Settlement.* 8vo, numerous Illustrations, 21s.
Book of the Play. By DUTTON COOK. New and Revised
Edition. 1 vol., cloth extra, 3s. 6d.
Bower (G. S.) Law relating to Electric Lighting. Crown
8vo, 5s.
Boy's Froissart (The). Selected from the Chronicles of
England, France, and Spain. Illustrated, square crown 8vo, 7s. 6d.
See "Froissart."
Boy's King Arthur (The). With very fine Illustrations.
Square crown 8vo, cloth extra, gilt edges, 7s. 6d. Edited by SIDNEY
LANIER, Editor of "The Boy's Froissart."
*Boy's Mabinogion (The): being the Original Welsh Legends of
King Arthur.* Edited by SIDNEY LANIER. With numerous very
graphic Illustrations. Crown 8vo, cloth, gilt edges, 7s. 6d.
Brassey (Lady) Tahiti. With Photos. by Colonel Stuart-
Wortley. Fcap. 4to, 21s.

Breton Folk: An Artistic Tour in Brittany. By HENRY BLACKBURN, Author of "Artists and Arabs," "Normandy Picturesque," &c. With 171 Illustrations by RANDOLPH CALDECOTT. Imperial 8vo, cloth extra, gilt edges, 21*s.*; plainer binding, 10*s.* 6*d.*

Bryant (W. C.) and Gay (S. H.) History of the United States. 4 vols., royal 8vo, profusely Illustrated, 60*s.*

Bryce (Prof.) Manitoba. Crown 8vo, 7*s.* 6*d.*

Burnaby (Capt.). See "On Horseback."

Burnham Beeches (Heath, F. G.). With numerous Illustrations and a Map. Crown 8vo, cloth, gilt edges, 3*s.* 6*d.* Second Edition.

Butler (W. F.) The Great Lone Land; an Account of the Red River Expedition, 1869-70. With Illustrations and Map. Fifth and Cheaper Edition, crown 8vo, cloth extra, 7*s.* 6*d.*

────── *Invasion of England, told twenty years after, by an Old* Soldier. Crown 8vo, 2*s.* 6*d.*

────── *The Wild North Land; the Story of a Winter Journey* with Dogs across Northern North America. Demy 8vo, cloth, with numerous Woodcuts and a Map, 4th Edition, 18*s.* Cr. 8vo, 7*s.* 6*d.*

────── *Red Cloud; or, the Solitary Sioux.* Imperial 16mo, numerous illustrations, gilt edges, 7*s.* 6*d.*

Buxton (H. J. W.) Painting, English and American. Crown 8vo, 5*s.*

CADOGAN *(Lady A.) Illustrated Games of Patience.* Twenty-four Diagrams in Colours, with Descriptive Text. Foolscap 4to, cloth extra, gilt edges, 3rd Edition, 12*s.* 6*d.*

California. Illustrated, 12*s.* 6*d.* See "Nordhoff."

Cambridge Trifles; or, Splutterings from an Undergraduate Pen. By the Author of "A Day of my Life at Eton," &c. 16mo, cloth extra, 2*s.* 6*d.*

Capello (H.) and Ivens (R.) From Benguella to the Territory of Yacca. Translated by ALFRED ELWES. With Maps and over 130 full-page and text Engravings. 2 vols., 8vo, 42*s.*

Carlyle (T.) Reminiscences of my Irish Journey in 1849. Crown 8vo, 7*s.* 6*d.*

Challamel (M. A.) History of Fashion in France. With 21 Plates, specially coloured by hand, satin-wood binding, imperial 8vo, 28*s.*

Changed Cross (The), and other Religious Poems. 16mo, 2*s.* 6*d.*

Child of the Cavern (The); or, Strange Doings Underground. By JULES VERNE. Translated by W. H. G. KINGSTON. Numerous Illustrations. Sq. cr. 8vo, gilt edges, 7*s.* 6*d.*; cl., plain edges, 3*s.* 6*d.*

Choice Editions of Choice Books. 2s. 6d. each. Illustrated by C. W. COPE, R.A., T. CRESWICK, R.A., E. DUNCAN, BIRKET FOSTER, J. C. HORSLEY, A.R.A., G. HICKS, R. REDGRAVE, R.A., C. STONEHOUSE, F. TAYLER, G. THOMAS, H. J. TOWNSHEND, E. H. WEHNERT, HARRISON WEIR, &c.

Bloomfield's Farmer's Boy.	Milton's L'Allegro.
Campbell's Pleasures of Hope.	Poetry of Nature. Harrison Weir.
Coleridge's Ancient Mariner.	Rogers' (Sam.) Pleasures of Memory.
Goldsmith's Deserted Village.	Shakespeare's Songs and Sonnets.
Goldsmith's Vicar of Wakefield.	Tennyson's May Queen.
Gray's Elegy in a Churchyard.	Elizabethan Poets.
Keat's Eve of St. Agnes.	Wordsworth's Pastoral Poems.

"Such works are a glorious beatification for a poet."—*Athenæum.*

Christ in Song. By Dr. PHILIP SCHAFF. A New Edition, revised, cloth, gilt edges, 6s.

Confessions of a Frivolous Girl (The): A Novel of Fashionable Life. Edited by ROBERT GRANT. Crown 8vo, 6s. Paper boards, 1s.

Coote (W.) Wanderings South by East. Illustrated, 8vo, 21s.

Cornet of Horse (The): A Story for Boys. By G. A. HENTY. Crown 8vo, cloth extra, gilt edges, numerous graphic Illustrations, 5s.

Cripps the Carrier. 3rd Edition, 6s. *See* BLACKMORE.

Cruise of H.M.S. "Challenger" (The). By W. J. J. SPRY, R.N. With Route Map and many Illustrations. 6th Edition, demy 8vo, cloth, 18s. Cheap Edition, crown 8vo, some of the Illustrations, 7s. 6d.

Cruise of the Walnut Shell (The). An instructive and amusing Story, told in Rhyme, for Children. With 32 Coloured Plates. Square fancy boards, 5s.

D'ANVERS (N.) An Elementary History of Art. Crown 8vo, 10s. 6d.

—— *Elementary History of Music.* Crown 8vo, 2s. 6d.

Daughter (A) of Heth. By W. BLACK. Crown 8vo, 6s.

Day of My Life (A); or, Every-Day Experiences at Eton. By an ETON BOY, Author of "About Some Fellows." 16mo, cloth extra, 2s. 6d. 6th Thousand.

Decoration. Vol. II., folio, 6s. Vol. III., New Series, folio, 7s. 6d.

De Leon (E.) Egypt under its Khedives. With Map and Illustrations. Crown 8vo, 4s.

Dick Cheveley: his Fortunes and Misfortunes. By W. H. G. KINGSTON. 350 pp., square 16mo, and 22 full-page Illustrations. Cloth, gilt edges, 7s. 6d.; plainer binding, plain edges, 5s.

Dick Sands, the Boy Captain. By JULES VERNE. With nearly 100 Illustrations, cloth, gilt, 10s. 6d.; plain binding and plain edges, 5s.

Don Quixote, Wit and Wisdom of. By EMMA THOMPSON. Square fcap. 8vo, 3s. 6d.

Donnelly (F.) Atlantis in the Antediluvian World. Crown 8vo, 12s. 6d.

Dos Passos (J. R.) Law of Stockbrokers and Stock Exchanges. 8vo, 35s.

EGYPT. See "Senior," "De Leon," "Foreign Countries."

Eight Cousins. See ALCOTT.

Electric Lighting. A Comprehensive Treatise. By J. E. H. GORDON. 8vo, fully Illustrated. [*In preparation.*

Elementary History (An) of Art. Comprising Architecture, Sculpture, Painting, and the Applied Arts. By N. D'ANVERS. With a Preface by Professor ROGER SMITH. New Edition, illustrated with upwards of 200 Wood Engravings. Crown 8vo, strongly bound in cloth, price 10s. 6d.

Elementary History (An) of Music. Edited by OWEN J. DULLEA. Illustrated with Portraits of the most eminent Composers, and Engravings of the Musical Instruments of many Nations. Crown 8vo, cloth, 2s. 6d.

Elinor Dryden. By Mrs. MACQUOID. Crown 8vo, 6s.

Embroidery (Handbook of). Edited by LADY MARIAN ALFORD, and published by authority of the Royal School of Art Needlework. With 22 Coloured Plates, Designs, &c. Crown 8vo, 5s.

Emerson (R. W.) Life and Writings. Crown 8vo, 8s. 6d.

English Catalogue of Books. Vol. III., 1872—1880. Royal 8vo, half-morocco, 42s.

—— *Dramatists of To-day.* By W. ARCHER, M.A. Crown 8vo, 8s. 6d.

English Philosophers. Edited by E. B. IVAN MÜLLER, M.A.

A series intended to give a concise view of the works and lives of English thinkers. Crown 8vo volumes of 180 or 200 pp., price 3s. 6d. each.

Francis Bacon, by Thomas Fowler.
Hamilton, by W. H. S. Monck.
Hartley and James Mill, by G. S. Bower.
*John Stuart Mill, by Miss Helen Taylor.
Shaftesbury and Hutcheson, by Professor Fowler.
Adam Smith, by J. A. Farrer.

* *Not yet published.*

Episodes in the Life of an Indian Chaplain. Crown 8vo, cloth extra, 12s. 6d.

Episodes of French History. Edited, with Notes, Maps, and Illustrations, by GUSTAVE MASSON, B.A. Small 8vo, 2s. 6d. each.
1. Charlemagne and the Carlovingians.
2. Louis XI. and the Crusades.
3. Part I. Francis I. and Charles V.
  „ II. Francis I. and the Renaissance.
4. Henry IV. and the End of the Wars of Religion.

Erema; or, My Father's Sin. 6s. *See* BLACKMORE.

Etcher (The). Containing 36 Examples of the Original Etched-work of Celebrated Artists, amongst others: BIRKET FOSTER, J. E. HODGSON, R.A., COLIN HUNTER, J. P. HESELTINE, ROBERT W. MACBETH, R. S. CHATTOCK, &c. Vols. for 1881 and 1882, imperial 4to, cloth extra, gilt edges, 2l. 12s. 6d. each.

Eton. See "Day of my Life," "Out of School," "About Some Fellows."

FARM Ballads. By WILL CARLETON. Boards, 1s.; cloth, gilt edges, 1s. 6d.

Farm Festivals. By the same Author. Uniform with above.

Farm Legends. By the same Author. See above.

Fashion (History of). See "Challamel."

Fechner (G. T.) On Life after Death. 12mo, vellum, 2s. 6d.

Felkin (R. W.) and Wilson (Rev. C. T.) Uganda and the Egyptian Soudan. An Account of Travel in Eastern and Equatorial Africa; including a Residence of Two Years at the Court of King Mtesa, and a Description of the Slave Districts of Bahr-el-Ghazel and Darfour. With a New Map of 1200 miles in these Provinces; numerous Illustrations, and Notes. By R. W. FELKIN, F.R.G.S., &c., &c.; and the Rev. C. T. WILSON, M.A. Oxon., F.R.G.S. 2 vols., crown 8vo, cloth, 28s.

Fern Paradise (The): A Plea for the Culture of Ferns. By F. G. HEATH. New Edition, fully Illustrated, large post 8vo, cloth, gilt edges, 12s. 6d. Sixth Edition.

Fern World (The). By F. G. HEATH. Illustrated by Twelve Coloured Plates, giving complete Figures (Sixty-four in all) of every Species of British Fern, printed from Nature; by several full-page and other Engravings. Cloth, gilt edges, 6th Edition, 12s. 6d.

Few Hints on Proving Wills (A). Enlarged Edition, 1s.

Fields (J. T.) Yesterdays with Authors. New Ed., 8vo., 16s.

First Steps in Conversational French Grammar. By F. JULIEN.
Being an Introduction to "Petites Leçons de Conversation et de Grammaire," by the same Author. Fcap. 8vo, 128 pp., 1s.

Florence. See "Yriarte."

Flowers of Shakespeare. 32 beautifully Coloured Plates. 5s.

Four Lectures on Electric Induction. Delivered at the Royal Institution, 1878-9. By J. E. H. GORDON, B.A. Cantab. With numerous Illustrations. Cloth limp, square 16mo, 3s.

Foreign Countries and British Colonies. A series of Descriptive Handbooks. Each volume will be the work of a writer who has special acquaintance with the subject. Crown 8vo, 3s. 6d. each.

Australia, by J. F. Vesey Fitzgerald.
Austria, by D. Kay, F.R.G.S.
*Canada, by W. Fraser Rae.
Denmark and Iceland, by E. C. Otté.
Egypt, by S. Lane Poole, B.A.
France, by Miss M. Roberts.
Greece, by L. Sergeant, B.A.
*Holland, by R. L. Poole.
Japan, by S. Mossman.
*New Zealand.
*Persia, by Major-Gen. Sir F. Goldsmid.
Peru, by Clements R. Markham, C.B.
Russia, by W. R. Morfill, M.A.
Spain, by Rev. Wentworth Webster.
Sweden and Norway, by F. H. Woods.
*Switzerland, by W. A. P. Coolidg M.A.
*Turkey-in-Asia, by J. C. McCoan, M.P.
West Indies, by C. H. Eden, F.R.G.S.

* *Not ready yet.*

Franc (Maud Jeanne). The following form one Series, small post 8vo, in uniform cloth bindings, with gilt edges:—

Emily's Choice. 5s.
Hall's Vineyard. 4s.
John's Wife: A Story of Life in South Australia. 4s.
Marian; or, The Light of Some One's Home. 5s.
Silken Cords and Iron Fetters. 4s.
Vermont Vale. 5s.
Minnie's Mission. 4s.
Little Mercy. 5s.
Beatrice Melton's Discipline. 4s.
No Longer a Child. 4s.
Golden Gifts. 5s.
Two Sides to Every Question. 5s.

Francis (F.) War, Waves, and Wanderings, including a Cruise in the "Lancashire Witch." 2 vols., crown 8vo, cloth extra, 24s.

Froissart (The Boy's). Selected from the Chronicles of England, France, Spain, &c. By SIDNEY LANIER. The Volume is fully Illustrated, and uniform with "The Boy's King Arthur." Crown 8vo, cloth, 7s. 6d.

From Newfoundland to Manitoba; a Guide through Canada's Maritime, Mining, and Prairie Provinces. By W. FRASER RAE. Crown 8vo, with several Maps, 6s.

Games of Patience. See CADOGAN.

Gentle Life (Queen Edition). 2 vols. in 1, small 4to, 6s.

THE GENTLE LIFE SERIES.
Price 6s. each; or in calf extra, price 10s. 6d.; Smaller Edition, cloth extra, 2s. 6d.

The Gentle Life. Essays in aid of the Formation of Character of Gentlemen and Gentlewomen.

About in the World. Essays by Author of "The Gentle Life."

Like unto Christ. A New Translation of Thomas à Kempis' "De Imitatione Christi."

Familiar Words. An Index Verborum, or Quotation Handbook. 6s.

Essays by Montaigne. Edited and Annotated by the Author of "The Gentle Life."

The Gentle Life. 2nd Series.

The Silent Hour: Essays, Original and Selected. By the Author of "The Gentle Life."

Half-Length Portraits. Short Studies of Notable Persons. By J. HAIN FRISWELL.

Essays on English Writers, for the Self-improvement of Students in English Literature.

Other People's Windows. By J. HAIN FRISWELL.

A Man's Thoughts. By J. HAIN FRISWELL.

Gilder (W. H.) Schwatka's Search. Sledging in quest of the Franklin Records. Illustrated, 8vo, 12s. 6d.

Gilpin's Forest Scenery. Edited by F. G. HEATH. Large post 8vo, with numerous Illustrations. Uniform with "The Fern World," re-issued, 7s. 6d.

Gordon (J. E. H.). See "Four Lectures on Electric Induction," "Physical Treatise on Electricity," "Electric Lighting."

Gouffé. The Royal Cookery Book. By JULES GOUFFÉ; translated and adapted for English use by ALPHONSE GOUFFÉ, Head Pastrycook to her Majesty the Queen. Illustrated with large plates printed in colours. 161 Woodcuts, 8vo, cloth extra, gilt edges, 2l. 2s.

—— Domestic Edition, half-bound, 10s. 6d.

"By far the ablest and most complete work on cookery that has ever been submitted to the gastronomical world."—*Pall Mall Gazette.*

Great Artists. See "Biographies."

Great Historic Galleries of England (The). Edited by LORD RONALD GOWER, F.S.A., Trustee of the National Portrait Gallery. Illustrated by 24 large and carefully executed *permanent* Photographs of some of the most celebrated Pictures by the Great Masters. Vol. I., imperial 4to, cloth extra, gilt edges, 36s. Vol. II., with 36 large permanent photographs, 2l. 12s. 6d.

Great Musicians. Edited by F. HUEFFER. A Series of Biographies, crown 8vo, 3s. each :—

Bach.	*Handel.	Schubert.
*Beethoven.	*Mendelssohn.	*Schumann.
*Berlioz.	*Mozart.	Richard Wagner.
English Church Composers.	Purcell.	Weber.
	Rossini.	

* *In preparation.*

Green (N.) A Thousand Years Hence. Crown 8vo, 6s.

Grohmann (W. A. B.) Camps in the Rockies. 8vo, 12s. 6d.

Guizot's History of France. Translated by ROBERT BLACK. Super-royal 8vo, very numerous Full-page and other Illustrations. In 8 vols., cloth extra, gilt, each 24s. This work is re-issued in cheaper binding, 8 vols., at 10s. 6d. each.

"It supplies a want which has long been felt, and ought to be in the hands of all students of history."—*Times.*

———————————— *Masson's School Edition.* The History of France from the Earliest Times to the Outbreak of the Revolution ; abridged from the Translation by Robert Black, M.A., with Chronological Index, Historical and Genealogical Tables, &c. By Professor GUSTAVE MASSON, B.A., Assistant Master at Harrow School. With 24 full-page Portraits, and many other Illustrations. 1 vol., demy 8vo, 600 pp., cloth extra, 10s. 6d.

Guizot's History of England. In 3 vols. of about 500 pp. each, containing 60 to 70 Full-page and other Illustrations, cloth extra, gilt, 24s. each ; re-issue in cheaper binding, 10s. 6d. each.

"For luxury of typography, plainness of print, and beauty of illustration, these volumes, of which but one has as yet appeared in English, will hold their own against any production of an age so luxurious as our own in everything, typography not excepted."—*Times.*

Guyon (Mde.) Life. By UPHAM. 6th Edition, crown 8vo, 6s.

HANDBOOK to the Charities of London. See LOW's.

Hall (W. W.) How to Live Long ; or, 1408 *Health Maxims,* Physical, Mental, and Moral. By W. W. HALL, A.M., M.D. Small post 8vo, cloth, 2s. 2nd Edition.

Harper's Monthly Magazine. Published Monthly. 160 pages, fully Illustrated. 1s.
>Vol. I. December, 1880, to May, 1881.
>,, II. May, 1881, to November, 1881.
>,, III. June to November, 1882.

Super-royal 8vo, 8s. 6d. each.

>"'Harper's Magazine' is so thickly sown with excellent illustrations that to count them would be a work of time; not that it is a picture magazine, for the engravings illustrate the text after the manner seen in some of our choicest *éditions de luxe*."—*St. James's Gazette.*
>
>"It is so pretty, so big, and so cheap.... An extraordinary shillingsworth— 160 large octavo pages, with over a score of articles, and more than three times as many illustrations."—*Edinburgh Daily Review.*
>
>"An amazing shillingsworth ... combining choice literature of both nations.'— *Nonconformist.*

Hatton (Joseph) Journalistic London: Portraits and En- gravings, with letterpress, of Distinguished Writers of the Day. Fcap. 4to, 12s. 6d.

—— *Three Recruits, and the Girls they left behind them.* Small post, 8vo, 6s.
>"It hurries us along in unflagging excitement."—*Times.*

Heart of Africa. Three Years' Travels and Adventures in the Unexplored Regions of Central Africa, from 1868 to 1871. By Dr. GEORG SCHWEINFURTH. Numerous Illustrations, and large Map. 2 vols., crown 8vo, cloth, 15s.

Heath (Francis George). See "Autumnal Leaves," "Burnham Beeches," "Fern Paradise," "Fern World," "Gilpin's Forest Scenery," "Our Woodland Trees," "Peasant Life," "Sylvan Spring," "Trees and Ferns," "Where to Find Ferns."

Heber's (Bishop) Illustrated Edition of Hymns. With upwards of 100 beautiful Engravings. Small 4to, handsomely bound, 7s. 6d. Morocco, 18s. 6d. and 21s. New and Cheaper Edition, cloth, 3s. 6d.

Heir of Kilfinnan (The). By W. H. G. KINGSTON. With Illustrations. Cloth, gilt edges, 7s. 6d.; plainer binding, plain edges, 5s.

Heldmann (Bernard) Mutiny on Board the Ship "Leander." Small post 8vo, gilt edges, numerous Illustrations, 7s. 6d.

Henty (G. A.) Winning his Spurs. Numerous Illustrations. Crown 8vo, 5s.

—— *Cornet of Horse;* which see.

Herrick (Robert) Poetry. Preface by AUSTIN DOBSON. With numerous Illustrations, by E. A. ABBEY. 4to, gilt edges, 42s.

History of a Crime (The); Deposition of an Eye-witness. The Story of the Coup d'État. By VICTOR HUGO. Crown 8vo, 6s.

History of Ancient Art. Translated from the German of JOHN WINCKELMANN, by JOHN LODGE, M.D. With very numerous Plates and Illustrations. 2 vols., 8vo, 36s.

——— *England.* See GUIZOT.

——— *English Literature.* See SCHERR.

——— *Fashion.* Coloured Plates. 28s. See CHALLAMEL.

——— *France.* See GUIZOT.

——— *Russia.* See RAMBAUD.

——— *Merchant Shipping.* See LINDSAY.

——— *United States.* See BRYANT.

History and Principles of Weaving by Hand and by Power. With several hundred Illustrations. By ALFRED BARLOW. Royal 8vo, cloth extra, 1l. 5s. Second Edition.

Hitchman (Francis) Public Life of the Right Hon: Benjamin Disraeli, Earl of Beaconsfield. New Edition, with Portrait. Crown 8vo, 3s. 6d.

Holmes (O. W.) The Poetical Works of Oliver Wendell Holmes. In 2 vols., 18mo, exquisitely printed, and chastely bound in limp cloth, gilt tops, 10s. 6d.

Hoppus (J. D.) Riverside Papers. 2 vols., 12s.

Hovgaard (A.) See "Nordenskiöld's Voyage." 8vo, 21s.

How I Crossed Africa: from the Atlantic to the Indian Ocean, Through Unknown Countries; Discovery of the Great Zambesi Affluents, &c.—Vol. I., The King's Rifle. Vol. II., The Coillard Family. By Major SERPA PINTO. With 24 full-page and 118 half-page and smaller Illustrations, 13 small Maps, and 1 large one. 2 vols., demy 8vo, cloth extra, 42s.

How to get Strong and how to Stay so. By WILLIAM BLAIKIE. A Manual of Rational, Physical, Gymnastic, and other Exercises. With Illustrations, small post 8vo, 5s.

Hugo (Victor) "*Ninety-Three.*" Illustrated. Crown 8vo, 6s.

——— *Toilers of the Sea.* Crown 8vo. Illustrated, 6s.; fancy boards, 2s.; cloth, 2s. 6d.; on large paper with all the original Illustrations, 10s. 6d.

——— *and his Times.* Translated from the French of A. BARBOU by ELLEN E. FREWER. 120 Illustrations, many of them from designs by Victor Hugo himself. Super-royal 8vo, cloth extra, 24s.

——— See "History of a Crime."

Hundred Greatest Men (The). 8 portfolios, 21s. each, or 4 vols., half-morocco, gilt edges, 12 guineas, containing 15 to 20 Portraits each. See below.

"Messrs. SAMPSON LOW & Co. are about to issue an important 'International' work, entitled, 'THE HUNDRED GREATEST MEN:' being the Lives and Portraits of the 100 Greatest Men of History, divided into Eight Classes, each Class to form a Monthly Quarto Volume. The Introductions to the volumes are to be written by recognized authorities on the different subjects, the English contributors being DEAN STANLEY, Mr. MATTHEW ARNOLD, Mr. FROUDE, and Professor MAX MÜLLER; in Germany, Professor HELMHOLTZ; in France, MM. TAINE and RENAN; and in America, Mr. EMERSON. The Portraits are to be Reproductions from fine and rare Steel Engravings."—*Academy.*

Hygiene and Public Health (A Treatise on). Edited by A. H. BUCK, M.D. Illustrated by numerous Wood Engravings. In 2 royal 8vo vols., cloth, One guinea each.

Hymnal Companion to Book of Common Prayer. See BICKERSTETH.

Illustrated Text-Books of Art-Education. Edited by EDWARD J. POYNTER, R.A. Each Volume contains numerous Illustrations, and is strongly bound for the use of Students, price 5s. The Volumes now ready are:—

PAINTING.

Classic and Italian. By PERCY R. HEAD.
German, Flemish, and Dutch.
French and Spanish.
English and American.

ARCHITECTURE.

Classic and Early Christian.
Gothic and Renaissance. By T. ROGER SMITH.

SCULPTURE.

Antique: Egyptian and Greek. | **Renaissance and Modern.**
Italian Sculptors of the 14th and 15th Centuries.

ORNAMENT.

Decoration in Colour. | **Architectural Ornament.**

Illustrated Dictionary (An) of Words used in Art and Archæology. Explaining Terms frequently used in Works on Architecture, Arms, Bronzes, Christian Art, Colour, Costume, Decoration, Devices, Emblems, Heraldry, Lace, Personal Ornaments, Pottery, Painting, Sculpture, &c., with their Derivations. By J. W. MOLLETT, B.A., Officier de l'Instruction Publique (France); Author of "Life of Rembrandt," &c. Illustrated with 600 Wood Engravings. Small 4to, strongly bound in cloth, 15s.

In my Indian Garden. By PHIL ROBINSON, Author of "Under the Punkah." With a Preface by EDWIN ARNOLD, M.A., C.S.I., &c. Crown 8vo, limp cloth, 4th Edition, 3s. 6d.

Irving (Washington). Complete Library Edition of his Works in 27 Vols., Copyright, Unabridged, and with the Author's Latest Revisions, called the "Geoffrey Crayon" Edition, handsomely printed in large square 8vo, on superfine laid paper, and each volume, of about 500 pages, will be fully Illustrated. 12s. 6d. per vol. *See also* "Little Britain."

——————— ("American Men of Letters.") 2s. 6d.

James (C.) Curiosities of Law and Lawyers. 8vo, 7s. 6d.

Johnson (O.) William Lloyd Garrison and his Times. Crown 8vo, 12s. 6d.

Jones (Major) The Emigrants' Friend. A Complete Guide to the United States. New Edition. 2s. 6d.

Kempis (Thomas à) Daily Text-Book. Square 16mo, 2s. 6d.; interleaved as a Birthday Book, 3s. 6d.

Kingston (W. H. G.). See "Snow-Shoes," "Child of the Cavern," "Two Supercargoes," "With Axe and Rifle," "Begum's Fortune," "Heir of Kilfinnan," "Dick Cheveley." Each vol., with very numerous Illustrations, square crown 16mo, gilt edges, 7s. 6d.; plainer binding, plain edges, 5s.

Lady Silverdale's Sweetheart. 6s. See BLACK.

Lanier. See "Boy's Froissart," "King Arthur," &c.

Lansdell (H.) Through Siberia. 2 vols., demy 8vo, 30s.; New Edition, very numerous illustrations, 8vo, 15s.

Larden (W.) School Course on Heat. Illustrated, crown 8vo, 5s.

Lathrop (G. P.) In the Distance. 2 vols., crown 8vo, 21s.

Lectures on Architecture. By E. VIOLLET-LE-DUC. Translated by BENJAMIN BUCKNALL, Architect. With 33 Steel Plates and 200 Wood Engravings. Super-royal 8vo, leather back, gilt top, with complete Index, 2 vols., 3l. 3s.

Leyland (R. W.) A Holiday in South Africa. Crown 8vo 12s. 6d.

Library of Religious Poetry. A Collection of the Best Poems of all Ages and Tongues. Edited by PHILIP SCHAFF, D.D., LL.D., and ARTHUR GILMAN, M.A. Royal 8vo, 1036 pp., cloth extra, gilt edges, 21*s.*; re-issue in cheaper binding, 10*s.* 6*d.*

Lindsay (W. S.) History of Merchant Shipping and Ancient Commerce. Over 150 Illustrations, Maps, and Charts. In 4 vols., demy 8vo, cloth extra. Vols. 1 and 2, 11*s.*; vols. 3 and 4, 14*s.* each. 4 vols. complete for 50*s.*

Little Britain; together with *The Spectre Bridegroom,* and *A Legend of Sleepy Hollow.* By WASHINGTON IRVING. An entirely New *Edition de luxe,* specially suitable for Presentation. Illustrated by 120 very fine Engravings on Wood, by Mr. J. D. COOPER. Designed by Mr. CHARLES O. MURRAY. Re-issue, square crown 8vo, cloth, 6*s.*

Long (Mrs. W. H. C.) Peace and War in the Transvaal. 12mo, 3*s.* 6*d.*

Lorna Doone. 6*s.*, 31*s.* 6*d.*, 35*s.* See "Blackmore."

Low's Select Novelets. Small post 8vo, cloth extra, 3*s.* 6*d.* each.
    Friends; a Duet. By E. S. PHELPS, Author of "The Gates Ajar."
    Baby Rue: Her Adventures and Misadventures, her Friends and her Enemies. By CHARLES M. CLAY.
    The Story of Helen Troy.
        "A pleasant book."—*Truth.*
    The Clients of Dr. Bernagius. From the French of LUCIEN BIART, by Mrs. CASHEL HOEY.
    The Undiscovered Country. By W. D. HOWELLS.
    A Gentleman of Leisure. By EDGAR FAWCETT.

Low's Standard Library of Travel and Adventure. Crown 8vo, bound uniformly in cloth extra, price 7*s.* 6*d.*, except where price is given.
  1. **The Great Lone Land.** By Major W. F. BUTLER, C.B.
  2. **The Wild North Land.** By Major W. F. BUTLER, C.B.
  3. **How I found Livingstone.** By H. M. STANLEY.
  4. **Through the Dark Continent.** By H. M. STANLEY. 12*s.* 6*d.*
  5. **The Threshold of the Unknown Region.** By C. R. MARKHAM. (4th Edition, with Additional Chapters, 10*s.* 6*d.*)
  6. **Cruise of the Challenger.** By W. J. J. SPRY, R.N.
  7. **Burnaby's On Horseback through Asia Minor.** 10*s.* 6*d.*
  8. **Schweinfurth's Heart of Africa.** 2 vols., 15*s.*
  9. **Marshall's Through America.**

Low's Standard Novels. Crown 8vo, 6s. each, cloth extra.

- **Work.** A Story of Experience. By LOUISA M. ALCOTT.
- **A Daughter of Heth.** By W. BLACK.
- **In Silk Attire.** By W. BLACK.
- **Kilmeny.** A Novel. By W. BLACK.
- **Lady Silverdale's Sweetheart.** By W. BLACK.
- **Sunrise.** By W. BLACK.
- **Three Feathers.** By WILLIAM BLACK.
- **Alice Lorraine.** By R. D. BLACKMORE.
- **Christowell, a Dartmoor Tale.** By R. D. BLACKMORE.
- **Clara Vaughan.** By R. D. BLACKMORE.
- **Cradock Nowell.** By R. D. BLACKMORE.
- **Cripps the Carrier.** By R. D. BLACKMORE.
- **Erema; or, My Father's Sin.** By R. D. BLACKMORE.
- **Lorna Doone.** By R. D. BLACKMORE.
- **Mary Anerley.** By R. D. BLACKMORE.
- **An English Squire.** By Miss COLERIDGE.
- **Mistress Judith.** A Cambridgeshire Story. By C. C. FRASER-TYTLER.
- **A Story of the Dragonnades; or, Asylum Christi.** By the Rev. E. GILLIAT, M.A.
- **A Laodicean.** By THOMAS HARDY.
- **Far from the Madding Crowd.** By THOMAS HARDY.
- **The Hand of Ethelberta.** By THOMAS HARDY.
- **The Trumpet Major.** By THOMAS HARDY.
- **Three Recruits.** By JOSEPH HATTON.
- **A Golden Sorrow.** By Mrs. CASHEL HOEY. New Edition.
- **Out of Court.** By Mrs. CASHEL HOEY.
- **History of a Crime:** The Story of the Coup d'État. VICTOR HUGO
- **Ninety-Three.** By VICTOR HUGO. Illustrated.
- **Adela Cathcart.** By GEORGE MAC DONALD.
- **Guild Court.** By GEORGE MAC DONALD.
- **Mary Marston.** By GEORGE MAC DONALD.
- **Stephen Archer.** New Edition of "Gifts." By GEORGE MAC DONALD.
- **The Vicar's Daughter.** By GEORGE MAC DONALD.
- **Weighed and Wanting.** By GEORGE MAC DONALD. [*In the Press.*
- **Diane.** By Mrs. MACQUOID.
- **Elinor Dryden.** By Mrs. MACQUOID.
- **My Lady Greensleeves.** By HELEN MATHERS.

Low's Standard Novels (continued):—
   John Holdsworth. By W. CLARK RUSSELL.
   A Sailor's Sweetheart. By W. CLARK RUSSELL.
   Wreck of the Grosvenor. By W. CLARK RUSSELL.
   The Afghan Knife. By R. A. STERNDALE.
   My Wife and I. By Mrs. BEECHER STOWE.
   Poganuc People, Their Loves and Lives. By Mrs. B. STOWE.
   Ben Hur: a Tale of the Christ. By LEW. WALLACE.

Low's Handbook to the Charities of London (Annual). Edited and revised to date by C. MACKESON, F.S.S., Editor of "A Guide to the Churches of London and its Suburbs," &c. Paper, 1s.; cloth, 1s. 6d.

MACDONALD (G.) *Orts.* Small post 8vo, 6s.

——— See also "Low's Standard Novels."

Macgregor (John) "Rob Roy" on the Baltic. 3rd Edition, small post 8vo, 2s. 6d.; cloth, gilt edges, 3s. 6d.

——— *A Thousand Miles in the "Rob Roy" Canoe.* 11th Edition, small post 8vo, 2s. 6d.; cloth, gilt edges, 3s. 6d.

——— *Description of the "Rob Roy" Canoe,* with Plans, &c., 1s.

——— *The Voyage Alone in the Yawl "Rob Roy."* New Edition, thoroughly revised, with additions, small post 8vo, 5s.; boards, 2s. 6d.

Macquoid (Mrs.). See LOW'S STANDARD NOVELS.

Magazine. See HARPER, UNION JACK, THE ETCHER, MEN OF MARK.

Magyarland. A Narrative of Travels through the Snowy Carpathians, and Great Alföld of the Magyar. By a Fellow of the Carpathian Society (Diploma of 1881), and Author of " The Indian Alps." 2 vols., 8vo, cloth extra, with about 120 Woodcuts from the Author's own sketches and drawings, 38s.

Manitoba: its History, Growth, and Present Position. By the Rev. Professor BRYCE, Principal of Manitoba College, Winnipeg. Crown 8vo, with Illustrations and Maps, 7s. 6d.

Markham (C. R.) The Threshold of the Unknown Region. Crown 8vo, with Four Maps, 4th Edition. Cloth extra, 10s. 6d.

Markham (C. R.) War between Peru and Chili, 1879-1881.
Crown 8vo, with four Maps, &c. [*In preparation.*]

Marshall (W. G.) Through America. New Edition, crown 8vo, with about 100 Illustrations, 7s. 6d.

Martin (J. W.) Float Fishing and Spinning in the Nottingham Style. Crown 8vo, 2s. 6d.

Marvin (Charles) The Russian Advance towards India. 8vo, 16s.

Maury (Commander) Physical Geography of the Sea, and its Meteorology. Being a Reconstruction and Enlargement of his former Work, with Charts and Diagrams. New Edition, crown 8vo, 6s.

Memoirs of Madame de Rémusat, 1802—1808. By her Grandson, M. PAUL DE RÉMUSAT, Senator. Translated by Mrs. CASHEL HOEY and Mr. JOHN LILLIE. 4th Edition, cloth extra. This work was written by Madame de Rémusat during the time she was living on the most intimate terms with the Empress Josephine, and is full of revelations respecting the private life of Bonaparte, and of men and politics of the first years of the century. Revelations which have already created a great sensation in Paris. 8vo, 2 vols., 32s.

—— *See also* "Selection."

Ménus (366, one for each day of the year). Each Ménu is given in French and English, with the recipe for making every dish mentioned. Translated from the French of COUNT BRISSE, by Mrs. MATTHEW CLARKE. Crown 8vo, 5s.

Men of Mark: a Gallery of Contemporary Portraits of the most Eminent Men of the Day taken from Life, especially for this publication, price 1s. 6d. monthly. Vols. I. to VII., handsomely bound, cloth, gilt edges, 25s. each.

Mendelssohn Family (The), 1729—1847. From Letters and Journals. Translated from the German of SEBASTIAN HENSEL. 3rd Edition, 2 vols., demy 8vo, 30s.

Michael Strogoff. See VERNE.

Mitford (Miss). See "Our Village."

Modern Etchings of Celebrated Paintings. 4to, 31s. 6d.

Mollett (J. W.) Illustrated Dictionary of Words used in Art and Archæology. Small 4to, 15s.

Morley (H.) English Literature in the Reign of Victoria. The 2000th volume of the Tauchnitz Collection of Authors. 18mo, 2s. 6d.

Music. See "Great Musicians."

NARRATIVES of State Trials in the Nineteenth Century.
First Period: From the Union with Ireland to the Death of George IV., 1801—1830. By G. LATHOM BROWNE, of the Middle Temple, Barrister-at-Law. 2nd Edition, 2 vols., crown 8vo, cloth, 26s.

Nature and Functions of Art (The); and more especially of Architecture. By LEOPOLD EIDLITZ. Medium 8vo, cloth, 21s.

Naval Brigade in South Africa (The). By HENRY F. NORBURY, C.B., R.N. Crown 8vo, cloth extra, 10s. 6d.

New Child's Play (A). Sixteen Drawings by E. V. B. Beautifully printed in colours, 4to, cloth extra, 12s. 6d.

Newfoundland. By FRASER RAE. See "From Newfoundland."

New Novels. Crown 8vo, cloth, 10s. 6d. per vol. :—
  The Granvilles. By the Hon. E. TALBOT. 3 vols.
  One of Us. By E. RANDOLPH.
  Weighed and Wanting. By GEORGE MAC DONALD. 3 vols.
  Castle Warlock. By GEORGE MAC DONALD. 3 vols.
  Under the Downs. By E. GILLIAT. 3 vols.
  A Stranger in a Strange Land. By LADY CLAY. 3 vols.
  The Heart of Erin. By Miss OWENS BLACKBURN. 3 vols.
  A Chelsea Householder. 3 vols.
  Two on a Tower. By THOMAS HARDY. 3 vols.
  The Lady Maud. By W. CLARK RUSSELL. 3 vols.

Nice and Her Neighbours. By the Rev. CANON HOLE, Author of "A Book about Roses," "A Little Tour in Ireland," &c. Small 4to, with numerous choice Illustrations, 12s. 6d.

Noah's Ark. A Contribution to the Study of Unnatural History. By PHIL ROBINSON. Small post 8vo, 12s. 6d.

Noble Words and Noble Deeds. From the French of E. MULLER. Containing many Full-page Illustrations by PHILIPPOTEAUX. Square imperial 16mo, cloth extra, 7s. 6d.; plainer binding, plain edges, 5s.

Nordenskiöld's Voyage around Asia and Europe. A Popular Account of the North-East Passage of the "Vega." By Lieut. A. HOVGAARD, of the Royal Danish Navy, and member of the "Vega" Expedition. 8vo, with about 50 Illustrations and 3 Maps, 21s.

Nordhoff (C.) California, for Health, Pleasure, and Residence. New Edition, 8vo, with Maps and Illustrations, 12s. 6d.

Nothing to Wear; and Two Millions. By W. A. BUTLER. New Edition. Small post 8vo, in stiff coloured wrapper, 1s.

Nursery Playmates (Prince of). 217 Coloured Pictures for Children by eminent Artists. Folio, in coloured boards, 6s.

Off to the Wilds: A Story for Boys. By G. MANVILLE FENN. Profusely Illustrated. Crown 8vo, 7s. 6d.

Old-Fashioned Girl. See ALCOTT.

On Horseback through Asia Minor. By Capt. FRED BURNABY. 2 vols., 8vo, 38s. Cheaper Edition, crown 8vo, 10s. 6d.

Our Little Ones in Heaven. Edited by the Rev. H. ROBBINS. With Frontispiece after Sir JOSHUA REYNOLDS. Fcap., cloth extra, New Edition—the 3rd, with Illustrations, 5s.

Our Village. By MARY RUSSELL MITFORD. Illustrated with Frontispiece Steel Engraving, and 12 full-page and 157 smaller Cuts. Crown 4to, cloth, gilt edges, 21s.; cheaper binding, 10s. 6d.

Our Woodland Trees. By F. G. HEATH. Large post 8vo, cloth, gilt edges, uniform with "Fern World" and "Fern Paradise," by the same Author. 8 Coloured Plates (showing leaves of every British Tree) and 20 Woodcuts, cloth, gilt edges, 12s. 6d. New Edition. About 600 pages.

Outlines of Ornament in all Styles. A Work of Reference for the Architect, Art Manufacturer, Decorative Artist, and Practical Painter. By W. and G. A. AUDSLEY, Fellows of the Royal Institute of British Architects. Only a limited number have been printed and the stones destroyed. Small folio, 60 plates, with introductory text, cloth gilt, 31s. 6d.

PALLISER (Mrs.) A History of Lace, from the Earliest Period. A New and Revised Edition, with additional cuts and text, upwards of 100 Illustrations and coloured Designs. 1 vol., 8vo, 1l. 1s.

—— *Historic Devices, Badges, and War Cries.* 8vo, 1l. 1s.

—— *The China Collector's Pocket Companion.* With upwards of 1000 Illustrations of Marks and Monograms. 2nd Edition, with Additions. Small post 8vo, limp cloth, 5s.

Pathways of Palestine: a Descriptive Tour through the Holy Land. By the Rev. CANON TRISTRAM. Illustrated with 44 permanent Photographs. (The Photographs are large, and most perfect Specimens of the Art.) Vols. I. and II., folio, gilt edges, 31s. 6d. each.

Peasant Life in the West of England. By FRANCIS GEORGE HEATH, Author of "Sylvan Spring," "The Fern World." Crown 8vo, 400 pp. (with Facsimile of Autograph Letter from Lord Beaconsfield to the Author, written December 28, 1880), 10s. 6d.

Petites Leçons de Conversation et de Grammaire: Oral and Conversational Method; the most Useful Topics of Conversation. By F. JULIEN. Cloth, 3s. 6d.

Photography (History and Handbook of). See TISSANDIER.

Physical Treatise on Electricity and Magnetism. By J. E. H. GORDON, B.A. With about 200 coloured, full-page, and other Illustrations. 2 vols., 8vo. New Edition. [*In preparation.*

Poems of the Inner Life. Chiefly from Modern Authors. Small 8vo, 5s.

Poganuc People: their Loves and Lives. By Mrs. BEECHER STOWE. Crown 8vo, cloth, 6s.

Polar Expeditions. See KOLDEWEY, MARKHAM, MACGAHAN, NARES, and NORDENSKIÖLD.

Poynter (Edward J., R.A.). See "Illustrated Text-books."

Prudence: a Story of Æsthetic London. By LUCY E. LILLIE. Small 8vo, 5s.

Publishers' Circular (The), and General Record of British and Foreign Literature. Published on the 1st and 15th of every Month, 3d.

Pyrenees (The). By HENRY BLACKBURN. With 100 Illustrations by GUSTAVE DORÉ, corrected to 1881. Crown 8vo, 7s. 6d.

RAE *(F.) Newfoundland.* See "From."

Redford (G.) Ancient Sculpture. Crown 8vo, 5s.

Reid (T. W.) Land of the Bey. Post 8vo, 10s. 6d.

Rémusat (Madame de). See "Memoirs of," "Selection."

Richter (Jean Paul). The Literary Works of Leonardo da Vinci. Containing his Writings on Painting, Sculpture, and Architecture, his Philosophical Maxims, Humorous Writings, and Miscellaneous Notes on Personal Events, on his Contemporaries, on Literature, &c.; for the first time published from Autograph Manuscripts. By J. P. RICHTER, Ph.Dr., Hon. Member of the Royal and Imperial Academy of Rome, &c. 2 vols., imperial 8vo, containing about 200 Drawings in Autotype Reproductions, and numerous other Illustrations. Price Eight Guineas to Subscribers. After publication the price will be Twelve Guineas.

—— *Italian Art in the National Gallery.* 4to.
[*Nearly ready.*

Robinson (Phil). See "In my Indian Garden," "Under the Punkah," "Noah's Ark," "Sinners and Saints."

Rose (J.) Complete Practical Machinist. New Edition, 12mo, 12s. 6d.

Rose Library (The). Popular Literature of all Countries. Each volume, 1s.; cloth, 2s. 6d. Many of the Volumes are Illustrated—
1. **Little Women.** By LOUISA M. ALCOTT.
2. **Little Women Wedded.** Forming a Sequel to "Little Women."
3. **Little Men.** By L. M. ALCOTT. Dble. vol., 2s.; cloth gilt, 3s. 6d.
4. **An Old-Fashioned Girl.** By LOUISA M. ALCOTT. Double vol., 2s.; cloth, 3s. 6d.
5. **Work.** A Story of Experience. By L. M. ALCOTT.
6. **Beginning Again.** Sequel to "Work." By L. M. ALCOTT.
7. **Stowe (Mrs. H. B.) The Pearl of Orr's Island.**
8. —— **The Minister's Wooing.**
9. —— **We and our Neighbours.** Double vol., 2s.; cloth, 3s. 6d.
10. —— **My Wife and I.** Double vol., 2s.; cloth, gilt 3s. 6d.
11. **Hans Brinker; or, the Silver Skates.** By Mrs. DODGE.
12. **My Study Windows.** By J. R. LOWELL.
13. **The Guardian Angel.** By OLIVER WENDELL HOLMES.
14. **My Summer in a Garden.** By C. D. WARNER.
15. **Dred.** Mrs. BEECHER STOWE. Dble. vol., 2s.; cloth gilt, 3s. 6d.
16. **Farm Ballads.** By WILL CARLETON.
17. **Farm Festivals.** By WILL CARLETON.
18. **Farm Legends.** By WILL CARLETON.
19, 20. **The Clients of Dr. Bernagius.** 2 parts, 1s. each.
21. **The Undiscovered Country.** By W. D. HOWELLS.
22. **Baby Rue.** By C. M. CLAY.

Round the Yule Log: Norwegian Folk and Fairy Tales. Translated from the Norwegian of P. CHR. ASBJÖRNSEN. With 100 Illustrations after drawings by Norwegian Artists, and an Introduction by E. W. Gosse. Imperial 16mo, cloth extra, gilt edges, 7s. 6d.

Rousselet (Louis) Son of the Constable of France. Small post 8vo, numerous Illustrations, 5s.

—— *The Drummer Boy: a Story of the Days of Washington.* Small post 8vo, numerous Illustrations, 5s.

Russell (W. Clark) The Lady Maud. 3 vols., crown 8vo, 31s. 6d.

—— See also LOW'S STANDARD NOVELS and WRECK.

Russell (W. H., LL.D.) Hesperothen: Notes from the Western World. A Record of a Ramble through part of the United States, Canada, and the Far West, in the Spring and Summer of 1881. By W. H. RUSSELL, LL.D. 2 vols., crown 8vo, cloth, 24s.

—— *The Tour of the Prince of Wales in India.* By W. H. RUSSELL, LL.D. Fully Illustrated by SYDNEY P. HALL, M.A. Super-royal 8vo, cloth extra, gilt edges, 52s. 6d.; Large Paper Edition, 84s.

Russian Literature. See "Turner."

SAINTS and their Symbols: A Companion in the Churches and Picture Galleries of Europe. With Illustrations. Royal 16mo, cloth extra, 3s. 6d.

Scherr (Prof. J.) History of English Literature. Translated from the German. Crown 8vo, 8s. 6d.

Schuyler (Eugène). The Life of Peter the Great. By EUGÈNE SCHUYLER, Author of "Turkestan." 2 vols., demy 8vo.
[*In preparation.*

Scott (Leader) Renaissance of Art in Italy. 4to, 31s. 6d.

Selection from the Letters of Madame de Rémusat to her Husband and Son, from 1804 to 1813. From the French, by Mrs. CASHEL HOEY and Mr. JOHN LILLIE. In 1 vol., demy 8vo (uniform with the "Memoirs of Madame de Rémusat," 2 vols.), cloth extra, 16s.

Senior (Nassau W.) Conversations and Journals in Egypt and Malta. 2 vols., 8vo, 24s.
These volumes contain conversations with SAID PASHA, ACHIM BEY, HEKEKYAN BEY, the Patriarch, M. DE LESSEPS, M. ST. HILAIRE, Sir FREDERICK BRUCE, Sir ADRIAN DINGLI, and many other remarkable people.

Seonee: Sporting in the Satpura Range of Central India, and in the Valley of the Nerbudda. By R. A. STERNDALE, F.R.G.S. 8vo, with numerous Illustrations, 21s.

Shadbolt (S.) The Afghan Campaigns of 1878—1880. By SYDNEY SHADBOLT, Joint Author of "The South African Campaign of 1879." 2 vols., royal quarto, cloth extra, 3l. 3s.

Shooting: its Appliances, Practice, and Purpose. By JAMES DALZIEL DOUGALL, F.S.A., F.Z.A., Author of "Scottish Field Sports," &c. New Edition, revised with additions. Crown 8vo, cloth extra, 7s. 6d.
"The book is admirable in every way..... We wish it every success."—*Globe.*
"A very complete treatise..... Likely to take high rank as an authority on shooting."—*Daily News.*

Sikes (Wirt). Rambles and Studies in Old South Wales. With numerous Illustrations. Demy 8vo, 18s.

Silent Hour (The). See "Gentle Life Series."

Silver Sockets (The); and other Shadows of Redemption. Eighteen Sermons preached in Christ Church, Hampstead, by the Rev. C. H. WALLER. Small post 8vo, cloth, 6s.

Sinners and Saints: a Tour across the United States of America, and Round them. By PHIL ROBINSON. [*In the Press.*

Sir Roger de Coverley. Re-imprinted from the "Spectator." With 125 Woodcuts, and steel Frontispiece specially designed and engraved for the Work. Small fcap. 4to, 6s.

Smith (G.) Assyrian Explorations and Discoveries. By the late GEORGE SMITH. Illustrated by Photographs and Woodcuts. Demy 8vo, 6th Edition, 18s.

—— *The Chaldean Account of Genesis.* By the late G. SMITH, of the Department of Oriental Antiquities, British Museum. With many Illustrations. Demy 8vo, cloth extra, 6th Edition, 16s. An entirely New Edition, completely revised and re-written by the Rev. PROFESSOR SAYCE, Queen's College, Oxford. Demy 8vo, 18s.

Smith (J. Moyr). See "Ancient Greek Female Costume."

Snow-Shoes and Canoes; or, the Adventures of a Fur-Hunter in the Hudson's Bay Territory. By W. H. G. KINGSTON. 2nd Edition. With numerous Illustrations. Square crown 8vo, cloth extra, gilt edges, 7s. 6d.; plainer binding, 5s.

South African Campaign, 1879 (*The*). Compiled by J. P. MACKINNON (formerly 72nd Highlanders), and S. H. SHADBOLT; and dedicated, by permission, to Field-Marshal H.R.H. The Duke of Cambridge. Containing a portrait and biography of every officer killed in the campaign. 4to, handsomely bound in cloth extra, 2l. 10s.

South Kensington Museum. Vol. II., 21s.

Stack (E.) Six Months in Persia. 2 vols., crown 8vo, 24s.

Stanley (H. M.) How I Found Livingstone. Crown 8vo, cloth extra, 7s. 6d.; large Paper Edition, 10s. 6d.

—— "*My Kalulu," Prince, King, and Slave.* A Story from Central Africa. Crown 8vo, about 430 pp., with numerous graphic Illustrations, after Original Designs by the Author. Cloth, 7s. 6d.

—— *Coomassie and Magdala.* A Story of Two British Campaigns in Africa. Demy 8vo, with Maps and Illustrations, 16s.

—— *Through the Dark Continent.* Cheaper Edition, crown 8vo, 12s. 6d.

State Trials. See "Narratives."

Stenhouse (Mrs.) An Englishwoman in Utah. Crown 8vo, 2s. 6d.

Stoker (Bram) Under the Sunset. Crown 8vo, 6s.

Story without an End. From the German of Carové, by the late Mrs. SARAH T. AUSTIN. Crown 4to, with 15 Exquisite Drawings by E. V. B., printed in Colours in Fac-simile of the original Water Colours; and numerous other Illustrations. New Edition, 7s. 6d.

—— square 4to, with Illustrations by HARVEY. 2s. 6d.

Stowe (Mrs. Beecher) Dred. Cheap Edition, boards, 2s. Cloth, gilt edges, 3s. 6d.

Stowe (Mrs Beecher) Footsteps of the Master. With Illustrations and red borders. Small post 8vo, cloth extra, 6s.

—— *Geography,* with 60 Illustrations. Square cloth, 4s. 6d.

—— *Little Foxes.* Cheap Edition, 1s.; Library Edition, 4s. 6d.

—— *Betty's Bright Idea.* 1s.

—— *My Wife and I; or, Harry Henderson's History.* Small post 8vo, cloth extra, 6s.*

—— *Minister's Wooing.* 5s.; Copyright Series, 1s. 6d.; cl., 2s.*

—— *Old Town Folk.* 6s.; Cheap Edition, 2s. 6d.

—— *Old Town Fireside Stories.* Cloth extra, 3s. 6d.

—— *Our Folks at Poganuc.* 6s.

—— *We and our Neighbours.* 1 vol., small post 8vo, 6s. Sequel to "My Wife and I."*

—— *Pink and White Tyranny.* Small post 8vo, 3s. 6d. Cheap Edition, 1s. 6d. and 2s.

—— *Queer Little People.* 1s.; cloth, 2s.

—— *Chimney Corner.* 1s.; cloth, 1s. 6d.

—— *The Pearl of Orr's Island.* Crown 8vo, 5s.*

—— *Woman in Sacred History.* Illustrated with 15 Chromo-lithographs and about 200 pages of Letterpress. Demy 4to, cloth extra, gilt edges, 25s.

Student's French Examiner. By F. JULIEN, Author of "Petites Leçons de Conversation et de Grammaire." Square cr. 8vo, cloth, 2s.

Studies in the Theory of Descent. By Dr. AUG. WEISMANN, Professor in the University of Freiburg. Translated and edited by RAPHAEL MELDOLA, F.C.S., Secretary of the Entomological Society of London. Part I.—"On the Seasonal Dimorphism of Butterflies," containing Original Communications by Mr. W. H. EDWARDS, of Coalburgh. With two Coloured Plates. Price of Part. I. (to Subscribers for the whole work only), 8s.; Part II. (6 coloured plates), 16s.; Part III., 6s. Complete, 2 vols., 40s.

Surgeon's Handbook on the Treatment of Wounded in War. By Dr. FRIEDRICH ESMARCH, Surgeon-General to the Prussian Army. Numerous Coloured Plates and Illustrations, 8vo, strongly bound, 1l. 8s.

* *See also* Rose Library.

Sylvan Spring. By FRANCIS GEORGE HEATH. Illustrated by 12 Coloured Plates, drawn by F. E. HULME, F.L.S., Artist and Author of "Familiar Wild Flowers;" by 16 full-page, and more than 100 other Wood Engravings. Large post 8vo, cloth, gilt edges, 12s. 6d.

TAHITI. By Lady BRASSEY, Author of the "Voyage of the Sunbeam." With 31 Autotype Illustrations after Photos. by Colonel STUART-WORTLEY. Fcap. 4to, very tastefully bound, 21s.

Taine (H. A.) "Les Origines de la France Contemporaine." Translated by JOHN DURAND.
    Vol. 1. **The Ancient Regime.** Demy 8vo, cloth, 16s.
    Vol. 2. **The French Revolution.** Vol. 1. do.
    Vol. 3. **Do.** do. Vol. 2. do.

Tauchnitz's English Editions of German Authors. Each volume, cloth flexible, 2s.; or sewed, 1s. 6d. (Catalogues post free on application.)

———— *(B.) German and English Dictionary.* Cloth, 1s. 6d.; roan, 2s.

———— *French and English Dictionary.* Paper, 1s. 6d.; cloth, 2s.; roan, 2s. 6d.

———— *Italian and English Dictionary.* Paper, 1s. 6d.; cloth, 2s.; roan, 2s. 6d.

———— *Spanish and English.* Paper, 1s. 6d.; cloth, 2s.; roan, 2s. 6d.

Taylor (W. M.) Paul the Missionary. Crown 8vo, 7s. 6d.

Thausing (Prof.) Preparation of Malt and the Fabrication of Beer. 8vo, 45s.

Thomas à Kempis. See "Birthday Book."

Thompson (Emma) Wit and Wisdom of Don Quixote. Fcap. 8vo, 3s. 6d.

Thoreau. By SANBORN. (American Men of Letters.) Crown 8vo, 2s. 6d.

Through America; or, Nine Months in the United States. By W. G. MARSHALL, M.A. With nearly 100 Woodcuts of Views of Utah country and the famous Yosemite Valley; The Giant Trees, New York, Niagara, San Francisco, &c.; containing a full account of Mormon Life, as noted by the Author during his visits to Salt Lake City in 1878 and 1879. Demy 8vo, 21s.; cheap edition, crown 8vo, 7s. 6d.

Through the Dark Continent: The Sources of the Nile; Around the Great Lakes, and down the Congo. By H. M. STANLEY. Cheap Edition, crown 8vo, with some of the Illustrations and Maps, 12s. 6d.

Through Siberia. By the Rev. HENRY LANSDELL. Illustrated with about 30 Engravings, 2 Route Maps, and Photograph of the Author, in Fish-skin Costume of the Gilyaks on the Lower Amur. 2 vols., demy 8vo, 30s. Cheaper Edition, 1 vol., 15s.

Tour of the Prince of Wales in India. See RUSSELL.

Trees and Ferns. By F. G. HEATH. Crown 8vo, cloth, gilt edges, with numerous Illustrations, 3s. 6d.
"A charming little volume."—*Land and Water.*

Tristram (Rev. Canon) Pathways of Palestine: A Descriptive Tour through the Holy Land. First Series. Illustrated by 44 Permanent Photographs. 2 vols., folio, cloth extra, gilt edges, 31s. 6d. each.

Turner (Edward) Studies in Russian Literature. (The Author is English Tutor in the University of St. Petersburgh.) Crown 8vo, 8s. 6d.

Two Supercargoes (The); or, Adventures in Savage Africa. By W. H. G. KINGSTON. Numerous Full-page Illustrations. Square imperial 16mo, cloth extra, gilt edges, 7s. 6d.; plainer binding, 5s.

UNDER the Punkah. By PHIL ROBINSON, Author of "In my Indian Garden." Crown 8vo, limp cloth, 5s.

Union Jack (The). Every Boy's Paper. Edited by G. A. HENTY and BERNARD HELDMANN. One Penny Weekly, Monthly 6d. Vol. I., New Series.

The Opening Numbers will contain:—
SERIAL STORIES.

Straight: Jack Archer's Way in the World. By G. A. HENTY.
Spiggott's School Days: A Tale of Dr. Merriman's. By CUTHBERT BEDE.
Sweet Flower; or, Red Skins and Pale Faces. By PERCY B. ST. JOHN.
Under the Meteor Flag. By HARRY COLLINGWOOD.
The White Tiger. By LOUIS BOUSSENARD. Illustrated.
A Couple of Scamps. By BERNARD HELDMANN.
Also a Serial Story by R. MOUNTNEY JEPHSON.

——— Vols. II. and III., 4to, 7s. 6d.; gilt edges, 8s.

VINCENT (F.) Norsk, Lapp, and Finn. By FRANK VINCENT, Jun., Author of "The Land of the White Elephant," "Through and Through the Tropics," &c. 8vo, cloth, with Frontispiece and Map, 12s.

Vivian (A. P.) Wanderings in the Western Land. 3rd Edition, 10s. 6d.

BOOKS BY JULES VERNE.

WORKS.	Large Crown 8vo. — Containing 350 to 600 pp. and from 50 to 100 full-page illustrations.		Containing the whole of the text with some illustrations.	
	In very handsome cloth binding, gilt edges.	In plainer binding, plain edges.	In cloth binding, gilt edges, smaller type.	Coloured Boards.
	s. d.	s. d.	s. d.	
Twenty Thousand Leagues under the Sea. Part I. Ditto Part II.	10 6	5 0	3 6	2 vols., 1s. each.
Hector Servadac	10 6	5 0	3 6	2 vols., 1s. each.
The Fur Country	10 6	5 0	3 6	2 vols., 1s. each.
From the Earth to the Moon and a Trip round it	10 6	5 0	2 vols., 2s. each.	2 vols., 1s. each.
Michael Strogoff, the Courier of the Czar	10 6	5 0	3 6	2 vols., 1s. each.
Dick Sands, the Boy Captain	10 6	5 0	3 6	2 vols., 1s. each.
Five Weeks in a Balloon	7 6	3 6	2 0	1s. 0d.
Adventures of Three Englishmen and Three Russians	7 6	3 6	2 0	1 0
Around the World in Eighty Days	7 6	3 6	2 0	1 0
A Floating City	7 6	3 6	2 0	1 0
The Blockade Runners			2 0	1 0
Dr. Ox's Experiment	7 6	3 6	2 0	1 0
Master Zacharius				
A Drama in the Air			2 0	1 0
A Winter amid the Ice				
The Survivors of the "Chancellor"	7 6	3 6	2 0	2 vols. 1s. each.
Martin Paz			2 0	1 0
The Mysterious Island, 3 vols.:—	22 6	10 6	6 0	3 0
Vol. I. Dropped from the Clouds	7 6	3 6	2 0	1 0
Vol. II. Abandoned	7 6	3 6	2 0	1 0
Vol. III. Secret of the Island	7 6	3 6	2 0	1 0
The Child of the Cavern	7 6	3 6	2 0	1 0
The Begum's Fortune	7 6	3 6		
The Tribulations of a Chinaman	7 6	3 6		
The Steam House, 2 vols.:—				
Vol. I. Demon of Cawnpore	7 6			
Vol. II. Tigers and Traitors	7 6			
The Giant Raft, 2 vols.:—				
Vol. I. Eight Hundred Leagues on the Amazon	7 6			
Vol. II. The Cryptogram	7 6			
Godfrey Morgan	7 6			

Celebrated Travels and Travellers. 3 vols. Demy 8vo, 600 pp., upwards of 100 full-page illustrations, 12s. 6d.; gilt edges, 14s. each:—
(1) The Exploration of the World.
(2) The Great Navigators of the Eighteenth Century.
(3) The Great Explorers of the Nineteenth Century.

WAITARUNA: *A Story of New Zealand Life.* By ALEXANDER BATHGATE, Author of "Colonial Experiences." Crown 8vo, cloth, 5s.

Waller (Rev. C. H.) The Names on the Gates of Pearl, and other Studies. By the Rev. C. H. WALLER, M.A. New Edition. Crown 8vo, cloth extra, 3s. 6d.

—— *A Grammar and Analytical Vocabulary of the Words in* the Greek Testament. Compiled from Brüder's Concordance. For the use of Divinity Students and Greek Testament Classes. By the Rev. C. H. WALLER, M.A. Part I. The Grammar. Small post 8vo, cloth, 2s. 6d. Part II. The Vocabulary, 2s. 6d.

—— *Adoption and the Covenant.* Some Thoughts on Confirmation. Super-royal 16mo, cloth limp, 2s. 6d.

—— *See also* "Silver Sockets."

Wanderings South by East: a Descriptive Record of Four Years of Travel in the less known Countries and Islands of the Southern and Eastern Hemispheres. By WALTER COOTE. 8vo, with very numerous Illustrations and a Map, 21s.

Warner (C. D.) Back-log Studies. Boards, 1s. 6d.; cloth, 2s.

—— *Mummies and Moslems.* 8vo, cloth, 12s.

Washington Irving's Little Britain. Square crown 8vo, 6s.

Weaving. See "History and Principles."

Webster. (American Men of Letters.) 18mo, 2s. 6d.

Weismann (A.) Studies in the Theory of Descent. 2 vols., 8vo, 40s.

Where to Find Ferns. By F. G. HEATH, Author of "The Fern World," &c.; with a Special Chapter on the Ferns round London; Lists of Fern Stations, and Descriptions of Ferns and Fern Habitats throughout the British Isles. Crown 8vo, cloth, price 3s.

White (Rhoda E.) From Infancy to Womanhood. A Book of Instruction for Young Mothers. Crown 8vo, cloth, 10s. 6d.

White (R. G.) England Without and Within. New Edition, crown 8vo, 10s. 6d.

Whittier (J. G.) The King's Missive, and later Poems. 18mo, choice parchment cover, 3s. 6d. This book contains all the Poems written by Mr. Whittier since the publication of "Hazel Blossoms."

—— *The Whittier Birthday Book.* Extracts from the Author's writings, with Portrait and numerous Illustrations. Uniform with the "Emerson Birthday Book." Square 16mo, very choice binding, 3s. 6d.

Wild Flowers of Switzerland. 17 Coloured Plates. 4to.
[*In preparation.*

Williams (H. W.) Diseases of the Eye. 8vo, 21*s.*

Wills, A Few Hints on Proving, without Professional Assistance. By a PROBATE COURT OFFICIAL. 5th Edition, revised with Forms of Wills, Residuary Accounts, &c. Fcap. 8vo, cloth limp, 1*s.*

Winks (W. E.) Lives of Illustrious Shoemakers. With eight Portraits. Crown 8vo, 7*s.* 6*d.*

With Axe and Rifle on the Western Prairies. By W. H. G. KINGSTON. With numerous Illustrations, square crown 8vo, cloth extra, gilt edges, 7*s.* 6*d.* ; plainer binding, 5*s.*

Woolsey (C. D., LL.D.) Introduction to the Study of International Law; designed as an Aid in Teaching and in Historical Studies. 5th Edition, demy 8vo, 18*s.*

Wreck of the Grosvenor. By W. CLARK RUSSELL, Author of "John Holdsworth, Chief Mate," "A Sailor's Sweetheart," &c. 6*s.* Third and Cheaper Edition.

Wright (the late Rev. Henry) The Friendship of God. With Biographical Preface by the Rev. E. H. BICKERSTETH, Portrait, &c. Crown 8vo, 6*s.*

YRIARTE (Charles) Florence: its History. Translated by C. B. PITMAN. Illustrated with 500 Engravings. Large imperial 4to, extra binding, gilt edges, 63*s.*

History; the Medici; the Humanists; letters; arts; the Renaissance; illustrious Florentines; Etruscan art; monuments; sculpture; painting.

www.ingramcontent.com/pod-product-compliance
Lightning Source LLC
Chambersburg PA
CBHW032133160426
43197CB00008B/631